THE POLYKETIDE METABOLITES

ELLIS HORWOOD BOOKS IN ORGANIC CHEMISTRY

Series Editor: Dr J. MELLOR, University of Southampton

Belen'kii, L.I.	**Chemistry of Organosulfur Compounds: General Problems**
Beyer, H. & Walter, W.	**Advanced Organic Chemistry**
Blazej, A. & Kosik, M.	**Phytomass as a Raw Material for Chemistry**
Brown, S.S. & Kodama, Y.	**Toxicology of Metals: Clinical and Experiment Research**
Corey, E.R., Corey, J.Y. & Gasper, P.P.	**Silicon Chemistry**
Dunn, A.D. & Rudorf, W.D.	**Carbon Disulphide in Organic Chemistry**
Foá, V., Emmet, E.A., Moroni, M. & Colonksi, A.	**Occupational and Environmental Chemical Hazards: Cellular and Biochemical Indices of Monitoring Toxicity**
Fresenius, P.	**Organic Chemical Nomenclature**
Hudlicky, M.	**Chemistry of Organic Fluorine Compounds: A Laboratory Manual 2nd (Revised) Edition**
Hudlicky, M.	**Reductions in Organic Chemistry**
Keese, R. & Muller, R.K.	**Fundamentals of Preparative Organic Chemistry**
Kova, J., Krutosikova, A. & Kada, R.	**Chemistry of Heterocyclic Compounds**
Lukevics, E., Pudova, O. & Sturkovich, R.	**Molecular Structure of Organosilicon Compounds**
Maksic, Z.B.	**Modelling of Structure and Properties of Molecules**
Manitto, P.	**Biosynthesis of Natural Products**
Mason, J.M.	**Practical Sonochemistry: User's Guide to Applications in Chemistry and Chemical Engineering**
Neidleman, S.L. & Geigert, J.	**Biohalogenation: Principles, Basic Roles and Applications**
O'Hagan, D.	**The Polyketide Metabolites**
Pizey, J.S.	**Synthetic reagents:**
	Volume 1: Dimethyl Formamide; Lithium Aluminium Hydride; Mercuric Oxide; Thionyl Chloride
	Volume 2: *N*-Bromosuccinide; Diazomethane; Manganese Dioxide; Raney Nickel
	Volume 3: Diborane; 2,3-Dichloro-5,6-Dicyanobenzoquinone DDQ; Iodine; Lead Tetra-acetate
	Volume 4: Mercuric Acetate; Periodic Acid and Periodates; Sulfuryl Chloride
	Volume 5: Thallium(III) Acetate and Trifluoroacetate; Ammonia; Iodine Monochloride
	Lithium Aluminium Hydride (reprinted from Synthetic Reagents Volume 1)
Sakurai, H.	**Organosilicon and Bioorganosilicon Chemistry: Structure, Bonding, Reactivity and Synthetic Application**
Smith, E.H.	**Comparative Oxidation of Organic Compounds**
Sterba, V. & Panchartek, J.	**Kinetic Methods in Organic Chemistry**
Trinajstic, N., Knop, J.V., Müller, W.R. & Szymanski, K.	**Computation Chemical Graph Theory: Generation of Structures by Computer Methods**
Tundo, P.	**Continuous Flow Methods in Organic Synthesis**
Vernin, G.	**The Chemistry of Heterocyclic Flavouring and Aroma Compounds**

THE POLYKETIDE METABOLITES

DAVID O'HAGAN B.Sc., Ph.D.
Department of Chemistry
University of Durham

ELLIS HORWOOD
NEW YORK LONDON TORONTO SYDNEY TOKYO SINGAPORE

First published in 1991 by
ELLIS HORWOOD LIMITED
Market Cross House, Cooper Street,
Chichester, West Sussex, PO19 1EB, England

A division of
Simon & Schuster International Group
A Paramount Communications Company

Every effort has been made to trace all copyright holders, but if any have been inadvertently overlooked, the publishers will be pleased to make the necessary arrangements at the earliest opportunity.

Typeset in Times by Ellis Horwood Limited
Printed and bound in Great Britain
by Hartnolls, Bodmin, Cornwall

British Library Cataloguing-in-Publication Data

O'Hagan, David
The polyketide metabolites. —
(Ellis Horwood series in organic chemistry)
I. Title. II. Series.
574.19
ISBN 0–13–683269–5

Library of Congress Cataloging-in-Publication Data available

Table of contents

Introduction

HISTORICAL PERSPECTIVE

The term *polyketide* was introduced into the chemical literature in 1907 when John Norman Collie of University College, London, presented a paper to the Chemical Society entitled 'Derivatives of the multiple ketene group' [1]. Collie proposed that ketene ($CH_2{=}C{=}O$) or its hydrolysis product, acetic acid, was the basic building block for a large number of aromatic plant metabolites, hence the designation, *polyketide*. He provided experimental observations in support of his hypothesis demonstrating that a diverse range of aromatics, including some simple naturally occurring species, could be generated from condensations of acetic acid, acetoacetate and higher homologues of acetate [2,3]. His ideas appear to have been formulated as early as 1893 [4] and he clearly understood that Nature would exploit facile chemical transformations for the biosynthesis of natural products;

> I think therefore, that when it is possible to show that by merely using the simplest kind of change, namely condensation, acetyl groups will produce pyridine, benzene and naphthalene derivatives, we have imitated nature in a remarkable manner.
>
> J. N. Collie *J. C. S.* 1893, **63**, 329.

Although Collie was essentially correct in retrospect, his polyketide analysis was limited and at the time considered something of an intellectual curiosity. As a result, it only re-emerged in the literature in the early 1950s after a greater body of structural and biosynthetic evidence had accumulated. At the same time he attempted to extend the analysis to account for the origin of sugars from the condensation of acetate units, but in this regard his speculations never developed. When Collie formally retired from his chair at University College in 1928, he was replaced by Sir Robert Robinson, but Collie remained active in the department as Emeritus Professor. Robinson held the post for only two years, but by all accounts the two became good friends, beyond their common professional interests both shared a passion for mountaineering and spent time together in this pursuit [5]. With this interaction Robinson clearly became aware of Collie's early work and views. During the mid 1940s Rittenberg and Bloch established the involvement of acetate in fatty acid biosynthesis by demonstrating the incorporation of isotopically labelled acetate into fatty acids [6,7]. Robinson, fortified and stimulated by these results, reaffirmed and extended Collie's polyketide hypothesis in 1948 [8]. He elaborated on these ideas more fully in the classic text *The structural relationships of natural products* in 1955, where he also predicted, essentially correctly, the origin of the tetracycline antibiotics and other polyketides [9]. Independently, in the early 1950s Arthur J. Birch began to outline his 'acetate hypothesis', with the idea that acetate was the building block of many aromatic plant metabolites. Birch recognized and acknowledged Collie's intuition in his papers [10,11] but was further stimulated by the contemporaneous developments that were emerging in fatty acid biosynthesis, and in particular the discovery led by Feodor Lynen that acetyl coenzyme-A acts as 'active acetate' in fatty acid and steroid construction [12,13]. Birch systematically analysed and confirmed the structures of many aromatic polyketide metabolites, demonstrating that they were compatible with a biosynthetic origin from the folding of extended β-keto chains derived from acetate [14,15]. The analysis was also extended to rationalize the origin of the plant flavonoids by acetate extension of a cinnamic acid derivative (see section 3.5.5). Credit must be given to Birch for testing his hypothesis experimentally. He was the first to carry out biosynthetic experiments on fungal aromatics [16], which placed the role of acetate beyond doubt. Significant too among his extensive contributions was the first experimental demonstration [17] that the common pendent methyl groups of fungal polyketides are derived from the methyl group of L-methionine (see section 3.3.1). The basic origin of the polyketides was now established. It is significant that developments in fatty acid biosynthesis accelerated the initial understanding of polyketide biosynthesis. There are obvious similarities between the two processes, and even today much of the stimulus and direction in probing polyketide biochemistry comes from developments in fatty acid biosynthesis.

STRUCTURE AND DIVERSITY

The polyketide metabolites are natural products which classically derive from the condensation of acetate units in a head to tail manner. Fatty acids are undoubtedly the most significant polyketide metabolites owing to their structural role in cellular membranes. However, a tremendous variety of polyketides exist, and all organisms from bacteria to man elaborate metabolites with a polyketide origin. The structural diversity of the polyketide metabolites is extensive, but they are unified on the basis of a common chemical event responsible for carbon–carbon bond formation during biosynthesis. This is a decarboxylative condensation of a malonate with an acylthioester to generate a β-keto-thioester, a reaction which is mediated by the *condensing enzyme* of polyketide synthases as shown in Fig. 1. This reaction

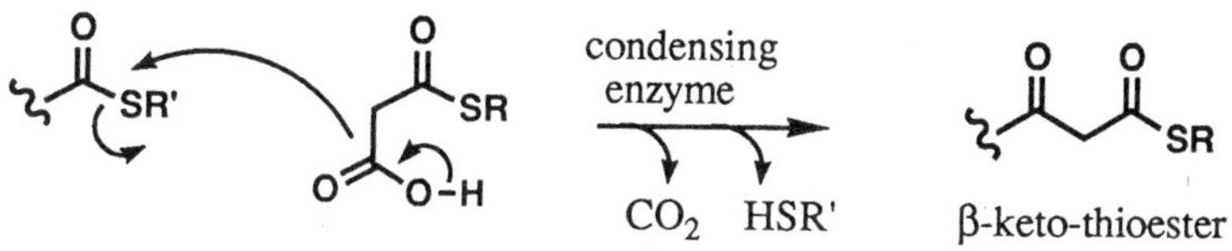

Fig. 1.

is repeated depending on the specification of each polyketide synthase, leading to the condensation of acetate groups, and can be viewed almost as a controlled oligomerization. The number of acetate groups condensed by a particular polyketide synthase is used to classify the polyketide as a triketide, tetraketide, pentaketide and so on. The carbon skeleton of all polyketide metabolites are assembled in this way, and it is the subsequent processing of the ketone functionality of the β-keto-thioester which is the key to the structural diversity. This processing can be viewed in three ways, two of which represent extremes and the third an intermediate level of manipulation, as shown in Fig. 2.

At one extreme the β-keto functionalities are completely reduced to methylene groups (route A in Fig. 2). This is the universal fatty acid pathway, which involves the reduction of the ketone to an alcohol followed by the elimination of the elements of water to generate an α,β-unsaturated double bond. The double bond is then reduced to generate the saturated methylene functionality (see section 1.1). At the other extreme (route D in Fig. 2), there is no reduction of the β-keto functionality and the polyketide metabolites arise from intramolecular condensations of extended poly-β-keto chains, usually ending up as aromatic phenols. This pathway is common in plants, fungi and bacteria, and many polyketide secondary metabolites, containing

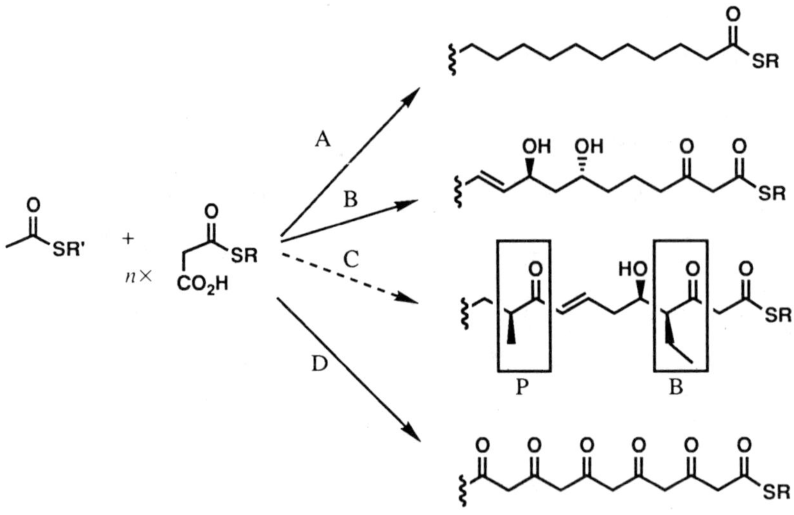

Fig. 2.

phenols as a partial structure, have an aromatic ring assembled in this way (see section 3.1). However, many polyketide secondary metabolites are in a continuum between these two extremes represented by route B. In these cases the polyketide synthases are programmed to reduce the β-ketones to varying degrees by employing *selectively* the reducing activities which are used in fatty acid biosynthesis. For example, the β-ketone can be reduced to an alcohol and remain as such, or the alcohol can be further eliminated to generate a double bond, or indeed the double bond can be completely reduced to a methylene group. Therefore, each β-ketone is adjusted to an individual status, leading to highly functionalized fatty acid derivatives which form the basis of the more complex polyketide metabolites (e.g. see sections 3.4 and 4.4.4). A further degree of complexity is found, particularly in the Actinomycete bacteria and marine organisms, when propionate and sometimes butyrate combine with acetate as subunits in polyketide assembly. The incorporation of propionate and butyrate leads to pendent methyl and ethyl groups respectively along the functionalized fatty acid derivatives, and the new chiral centres further increase the level of structural complexity (see route C in Fig. 2). The boxed subunits illustrate the incorporation of a propionate (P) and a butyrate (B) into the polyketide backbone. The macrolide and polyether antibiotics are such metabolites (see section 4.4.2), and many of the marine polyketides also incorporate propionate as a subunit (see section 5.4).

Polyketide metabolites elicit a wide variety of physiological effects and many compounds are exploited commercially, particularly when they are of microbial origin and are amenable to large scale production by fermentation. Of great significance in this regard are the numerous antibiotics of Actinomycetes which find application both in clinical and veterinary medicine (see Chapter 4). Clearly the desire to find new and improved pharmaceutical agents is the major stimulus for the continuous quest for and discovery of new polyketides, and in this respect their abundance and diversity is fortuitous. However, this diversity has not evolved to serve *our* needs. The existence of organisms depends on their ability or fitness to survive in hostile competitive environments and secondary metabolites are in the front line in this battle. The continuously changing pressures on organisms require versatile pathways for generating new defensive agents, and the polyketide pathway, like others leading, for example, to the terpenes and alkaloids, displays this versatility and can generate an infinite array of structures. Consequently, closely related organisms, in evolutionary terms, should elaborate secondary metabolites which are both biosynthetically and structurally similar. This book sets out to explore this by assessing and comparing the structures and modes of biosynthesis of polyketide metabolites from different classes of organism.

Often structural similarities between metabolites can suggest a misleading biosynthetic interpretation (e.g. radiclonic acid (see section 3.3) and viridcatumtoxin (see section 4.1)) and therefore biosynthetic information is invaluable in securing true biogenetic relationships; such studies underpin much of this analysis. Those structural correlations and relationships which have been suggested without biosynthetic evidence, and in certain instances on structure alone, however persuasive, are tentative. This is particularly pertinent for the marine metabolites discussed in Chapter 5 where an increasing number of structures are reported in the literature each year but very little biosynthetic information is forthcoming owing to the particular problems there are in working with marine organisms (see Chapter 5).

BIOSYNTHETIC METHODS

Most of the biosynthetic information on the polyketides has been obtained after incorporation of isotopically enriched precursors into the metabolites during their biosynthesis. The first biosynthetic studies carried out in the late 1950s and early 1960s utilized radioactively labelled precursors. Although revealing, these studies were labour intensive and time consuming as the metabolites had to be carved up chemically to isolate individual carbon atoms, such that the specific location of the radioactivity could be determined. Birch's study on griseofulvin [16] and Corcoran's on erythromycin [18] are elegant examples which required meticulous experimental technique and serve to illustrate that detailed and valuable information was obtained by such an approach. The situation has, however, changed. With the rapid developments in nuclear magnetic resonance (NMR) technology over the last twenty years there has been an explosion in the number of papers reporting biosynthetic studies on secondary metabolites using stable isotopes, and in particular employing ^{13}C-labelled precursors [19]. For polyketide metabolites the incorporation of [1-^{13}C]- and [2-^{13}C]-acetates and [*methyl*-^{13}C]-L-methionine are now routine experiments provided that the ^{13}C-NMR spectra of the metabolite under investigation can be correctly assigned. If a particular set of ^{13}C-NMR signals in the resultant spectrum are enriched over natural abundance (1.1%), then incorporation of the isotopically labelled carbon of the precursor into those enriched carbons of the metabolite can be deduced. The information is regiospecific and can be obtained directly without recourse to chemical degradation.

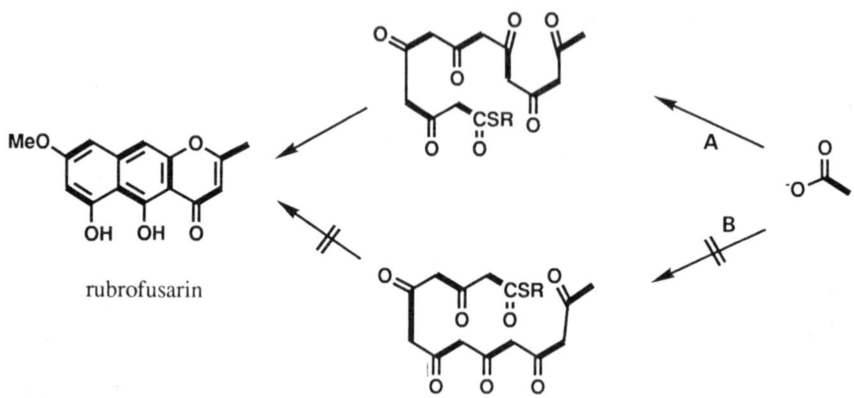

Fig. 3.

Double labelling studies employing [$^{13}C_2$]- acetate have served as a powerful means for determining the sites of incorporation of intact acetate units and in particular delineating the true polyketide folding pattern of a particular metabolite from several theoretical options. Rubrofusarin, a fungal metabolite of *Fusarium culmorum*, is an example where the labelling pattern from [1,2-$^{13}C_2$]-acetate demonstrated that the polyketide folded by pathway A and *not* by pathway B (Fig. 3) [20]. Two carbon-13 atoms sited adjacent to each other will couple to each other in the ^{13}C-NMR experiment and the persistence of coupled signals in the metabolite after incorporation of [$^{13}C_2$]-acetate suggests that the bond between the coupled carbons has remained intact during biosynthesis and that both the carbons are derived from the same acetate unit. Beyond these basic experiments, carbon-13 has been used extensively as a reporter nucleus to establish the integrity of intact ^{13}C–^{2}H, ^{13}C–^{12}C–^{2}H and ^{13}C–^{18}O bonds in going from doubly labelled precursors to natural products. The presence of the heavy isotopes ^{2}H and ^{18}O induces an observable upfield shift in the resultant ^{13}C-NMR spectrum when the heavy isotope is attached directly (α-shift) or one bond removed (β-shift) to the ^{13}C-atom [21,22]. For example, with the availability of [1-^{13}C,2-$^{2}H_3$]-, [2-^{13}C,2-$^{2}H_3$]- and [1-^{13}C, $^{18}O_2$]-acetates, experiments can be designed to probe the origin of *all* of the carbon, hydrogen and oxygen atoms of acetogenic metabolites, and information obtained on bonds which are formed and broken during the biosynthetic process. Additionally, organisms can be grown under an atmosphere of heavy oxygen, $^{18}O_2$, to determine the source of oxygen atoms from molecular oxygen (e.g. see section 4.4.9).

In certain cases blocked mutants have revealed a great deal of information about biosynthetic pathways. Typically a biosynthetic pathway involves a series of enzymatic steps where a substrate is modified and then passed onto the next enzyme for processing and so on until the metabolite is constructed. Mutants of the parent or wild type organism which are generated by UV or chemically induced mutagenesis can be selected which are devoid of the capacity to express one of the biosynthetic enzymes because the DNA coding that protein has been damaged. Therefore, the biosynthetic pathway will stop at this point and the substrate for the missing enzyme will accumulate and can be identified as an intermediate. With various mutants, blocked at different points on a biosynthetic pathway, cross-referencing experiments can be conducted. An advanced intermediate secreted by one mutant should be transformed into the end product by another mutant blocked in the capacity to synthesize that intermediate. In this way details of a biosynthetic pathway can be constructed and the point at which each mutant is blocked can be identified. Some of the details on the biosynthesis of the fungal aflatoxins (see section 3.2.6) and much of the detail of tetracycline antibiotic biosynthesis (see section 4.1) have been obtained from blocked mutant studies. The text in general does not describe the nature and particulars of individual biosynthetic experiments, but reports the results and conclusions of such studies, and in many instances specific labelling patterns are illustrated.

PRÉCIS

The mechanistic detail which has emerged from the collective biosynthetic data has revealed that fatty acid and polyketide assembly are closely analogous. Fatty acid biosynthesis has been studied extensively in different classes of organism and offers an excellent model by which to introduce the processes of polyketide assembly. The first chapter of this book therefore focuses on the structure and biosynthesis of fatty acids.

Prostanoids and related hormones, which are derived from unsaturated fatty acids, are discussed in the second chapter. The structure and biosynthesis of the mammalian prostaglandins, prostacyclin, the thromboxanes and the leukotrienes are highlighted. Additionally, this chapter reviews the occurrence of prostaglandins from other sources. It is well-known that prostaglandins occur in soft coral, but there is a body of evidence accumulating which suggests that they also play a regulatory role in the metabolism and reproduction of some insects (see section 2.2.2) and that they may be involved in the regulation of flowering in higher plants (see section 2.2.3).

The remaining chapters deal with secondary metabolites from fungal and plant, bacterial and marine sources. The theme, particularly through these final three chapters, is to demonstrate the variety of structures that exist and illustrate the versatility of the polyketide pathway. In general the polyketides of the fungi and plants are true acetogenins, i.e. they derive solely from acetate. They tend to have aromatic nuclei and are structurally less complex than those polyketides of bacteria and marine organisms. Some fungi do, however, elaborate reduced polyketides (see section 3.4), which have biosynthetic and structural characteristics similar to the bacterial metabolites, and reciprocally many of the bacterial polyketides are acetogenic aromatics (see sections 4.1–4.3). So there are clear areas of overlap. It is notable, however, that the fungi and plants rarely employ subunits other than acetate for biosynthesis, whereas the bacteria have evolved polyketide synthases which utilize both propionate and butyrate in addition to acetate. For example, the polyketide aglycone of the erythromycins is constructed entirely from propionate (see section 4.4.1). These complex metabolites carry much structural information, and comparisons can be made between different metabolites to assess areas of structural homology. The macrolide antibiotics are remarkable in this context. Around 80 structures have been identified, and despite their complex and varied functionality all possess a common and predictable absolute stereochemistry (see section 4.4.5). Clearly compounds with homologous structural and stereochemical features, such as the macrolides, must possess biogenetically related polyketide synthases. This chapter outlines the different classes of bacterial polyketides and highlights the structural relationships that exist within and between the various classes (see section 4.5). The genetic analysis of polyketide producers is providing information on the structure of polyketide synthases and how they programme the assembly of individual metabolites. Clues to the origin of the common structural and stereochemical features, which exist between the higher polyketides such as the macrolide antibiotics, are beginning to emerge from this analysis (see section 4.8.3).

The final chapter considers the polyketides of marine origin. These are a diverse group of metabolites produced by a diverse range of organisms. Some of the metabolites contain structural characteristics similar to the bacterial polyketides. The mollusks, for example, secrete an interesting variety of polypropionate metabolites (see section 5.4). Others are quite unique. The biosynthetic evidence from studies on the brevetoxins suggests that the planktonic dinoflagellates have evolved a modified polyketide pathway (see section 5.2). The essentially untapped potential of marine organisms as a source of novel biologically active compounds will ensure their continued investigation, and as more compounds are isolated and their producing organisms interrogated we can look forward to new and exciting discoveries in natural product chemistry.

REFERENCES

[1] J. N. Collie, Derivatives of the multiple ketene group. *Proc. Chem. Soc.*, 1907, **91**, 230.

[2] J. N. Collie and T. P. Hilditch, An isomeric change of dehydracetic acid. *J. Chem. Soc.*, 1907, **91**, 787.

[3] J. N. Collie, Derivatives of the multiple ketene group. *J. Chem. Soc.*, 1907, **91**, 1806.

[4] J. N. Collie, The production of naphthalene derivatives from dehydracetic acid. *J. Chem. Soc.*, 1893, **63**, 329.

[5] T. I. Williams, *Robert Robinson: chemist extraordinary*. Oxford University Press, 1990.

[6] D. Rittenberg and K. Bloch, The utilisation of AcOH for fatty acid synthesis. *J. Biol. Chem.*, 1944, **154**, 311.

[7] D. Rittenberg and K. Bloch, Utilisation of acetic acid for the synthesis of fatty acids. *J. Biol. Chem.*, 1945, **160**, 417.

[8] R. Robinson, Structural relations of some plant products. *J. Roy. Soc. Arts.*, 1948, **96**, 795.

[9] R. Robinson, *The structural relations of natural products*. Oxford University Press, 1955.

[10] A. J. Birch and F. W. Donovan, Studies in relation to biosynthesis I. Some possible routes to derivatives of orcinol and phloroglucinol. *Aust. J. Chem.*, 1953, **6**, 360.

[11] A. J. Birch, Biosynthesis of polyketide and related compounds. *Science*, 1967, **156**, 202.

[12] F. Lynen, E. Reichert and L. Rueff, Zum biologischen abbau der essigaure IV. 'Aktivierten essigature.' ihre isolierung aus hefe und ihre chemische natur. *Liebigs Ann. Chem.*, 1951, **574**, 1.

[13] F. Lynen and E. Reichert, Zur chemischen struktur der 'aktivierten essigaure'. *Agnew Chem.*, 1951, **63**, 47.

[14] A. J. Birch and P. Elliot, Studies in relation to biosynthesis II. The structure of macrophone. *Aust. J. Chem.*, 1953, **6**, 369.

[15] A. J. Birch and F. W. Donovan, Studies in relation to biosynthesis III. The structure of eleutherinol. *Aust. J. Chem.*, 1953, **6**, 373.

[16] A. J. Birch, R. A. Massey-Westropp, R. W. Rickards and H. Smith, Studies in relation to biosynthesis. Part XIII. Griseofulvin. *J. Chem. Soc.*, 1958, 360.

[17] A. J. Birch, R. J. English, R. A. Massey-Westropp, M. Slater and H. Smith, Studies in relation to biosynthesis Part XIV. The origin of the nuclear methyl groups in mycophenolic acid. *J. Chem. Soc.*, 1958, 365.

[18] T. Kaneda, J. C. Butte, S., B. Taubmann and J. W. Corcoran, Actinomycete antibiotics. The biogenesis of erythronolide, the C_{21} branched chain lactone in erythromycin. *J. Biol. Chem.*, 1962, **237**, 322.

[19] J. C. Vederas, The use of stable isotopes in biosynthesis. *Nat. Prod. Rep.*, 1987, **4**, 277.

[20] F. J. Leeper and J. Staunton, The biosynthesis of rubrofusarin, a polyketide naphthopyrone from *Fusarium culmorum*: ^{13}C NMR assignments and incorporation of ^{13}C- and ^{2}H-labelled acetates. *J. C. Soc. Perkin Trans. I.*, 1984, 2919.

[21] M. J. Garson and J. Staunton, Some new NMR methods for tracing the fate of hydrogen in biosynthesis. *Chem. Soc. Rev.*, 1979, 539.

[22] J. C. Vederas, Structural dependence of ^{18}O isotope shifts in ^{13}C NMR spectra. *J. Am. Chem. Soc.*, 1980, **102**, 374.

John Norman Collie 1859–1942.
(Photograph courtesy of Department of Chemistry,
University College, London)

1

The fatty acids

Fatty acids are primary metabolites common to almost all living cells† and are characterized by a long hydrocarbon chain with a carboxylic acid terminus. Nature has exploited their hydrophobic properties, which force them to aggregate in the aqueous environment of the cell, and the fatty acids, as constituents of phospholipids, play a critical structural role in the assembly of cell membranes. Fatty acids are found also in cells as triglycerides (fats) where they are essentially stored as a food reserve. The more common natural saturated and unsaturated fatty acids are detailed in Table 1.1.

Table 1.1 — The common fatty acids

	Common name (code)†	Systematic name	Structural formula
Saturated			
C_8	Caprylic	*n*-Octanoic	$CH_3(CH_2)_6CO_2H$
C_{10}	Capric	*n*-Decanoic	$CH_3(CH_2)_8CO_2H$
C_{12}	Lauric	*n*-Dodecanoic	$CH_3(CH_2)_{10}CO_2H$
C_{14}	Myristic	*n*-Tetradecanoic	$CH_3(CH_2)_{12}CO_2H$
C_{16}	Palmitic	*n*-Hexadecanoic	$CH_3(CH_2)_{14}CO_2H$
C_{18}	Stearic	*n*-Octadecanoic	$CH_3(CH_2)_{16}CO_2H$
C_{20}	Arachidic	*n*-Eicosanoic	$CH_3(CH_2)_{18}CO_2H$
Unsaturated			
C_{18}	Oleic (18:1,9*Z*)	*cis*-Octadeca-9-enoic	
C_{18}	Vaccenic (18:1,11*Z*)	*cxis*-Octadeca-11-enoic	
C_{18}	Linoleic (18:2,9*Z*,12*Z*)	Octadeca-9,12-dienoic	
C_{18}	Linolenic (18:3,9*Z*,12*Z*,15*Z*)	Octadeca-9,12,15-trienoic	
C_{20}	Arachidonic (20:4,5Z,8Z,11Z,14Z)	Eicosa-5,8,11,14,-tetraenoic	

†The code defines (the number of carbon atoms, the number of double bonds, the position of the double bond and geometry $Z = cis$ and $E = trans$).

The process by which saturated fatty acids are assembled offers a blueprint for the construction of all polyketide metabolites. Organisms biosynthesize fatty acids by an essentially universal pathway from acetyl

† The exception are the unusual archaebacteria which possess phytanyl ether lipids derived from the terpenoid pathway.

coenzyme-A (acetyl-CoA) (1.1), although there are significant variations in the way different organisms organize their cellular machinery to do this. *De Novo* fatty acid biosynthesis requires the sequential condensation and reduction of acyl thioesters. For the condensations to occur acetyl-CoA is primed by carboxylation to malonyl-CoA and then transesterified onto a fatty acid synthase (FAS). The key condensation events of the growing fatty acid chain are driven by decarboxylation, and a series of enzymatic transformations are then required to reduce the resultant β-ketone functionality all the way to saturated methylene groups. The saturated thioester is then at liberty to undergo a further condensation reaction with another malonyl species, followed by repetition of the reduction sequence, to extend the chain by a further two carbon atoms (Fig. 1.1).

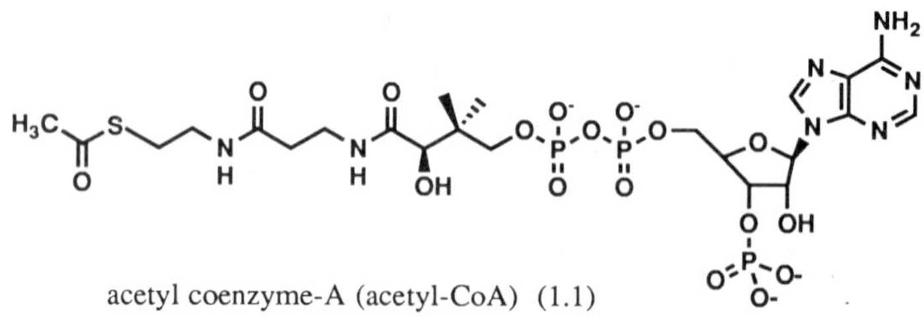

acetyl coenzyme-A (acetyl-CoA) (1.1)

FASs, particularly in higher organisms, are among the most sophisticated enzyme complexes that have been identified. Typically the acyl and malonyl thioesters are processed by the synthase and the fatty acid is assembled in its entirety without the release of partially constructed intermediates [1]. There is a wide diversity in the structure of FASs utilized by different organisms, extending from a more primitive collection of dissociated enzymes in bacteria and plants (Type II FAS) [2], to the mammalian fatty acid synthases, which possess all the necessary activities on one or two multifunctional proteins (Type I FAS). (See section 1.2.)

1.1 FATTY ACID BIOSYNTHESIS

Fatty acid biosynthesis can be defined in two steps. The first is the generation of malonyl-CoA by the action of acetyl-CoA carboxylase on acetyl-CoA, and the second is the conversion of acetyl-CoA and malonyl-CoA into the fatty acid by the FAS, in the presence of NADPH as a co-factor.

Step 1 $CH_3COSCoA + HCO_3^- + ATP \rightarrow HO_2CCH_2COSCoA + ADP + P_i$

Step 2 $CH_3COSCoA + nHO_2CCH_2COSCoA + 2nNADPH + 2nH^+ \rightarrow$
$CH_3CH_2(CH_2CH_2)_{n-1}CH_2CO_2H + 2nNADP^+ +$
$_{n+1}CoASH + nH_2O$

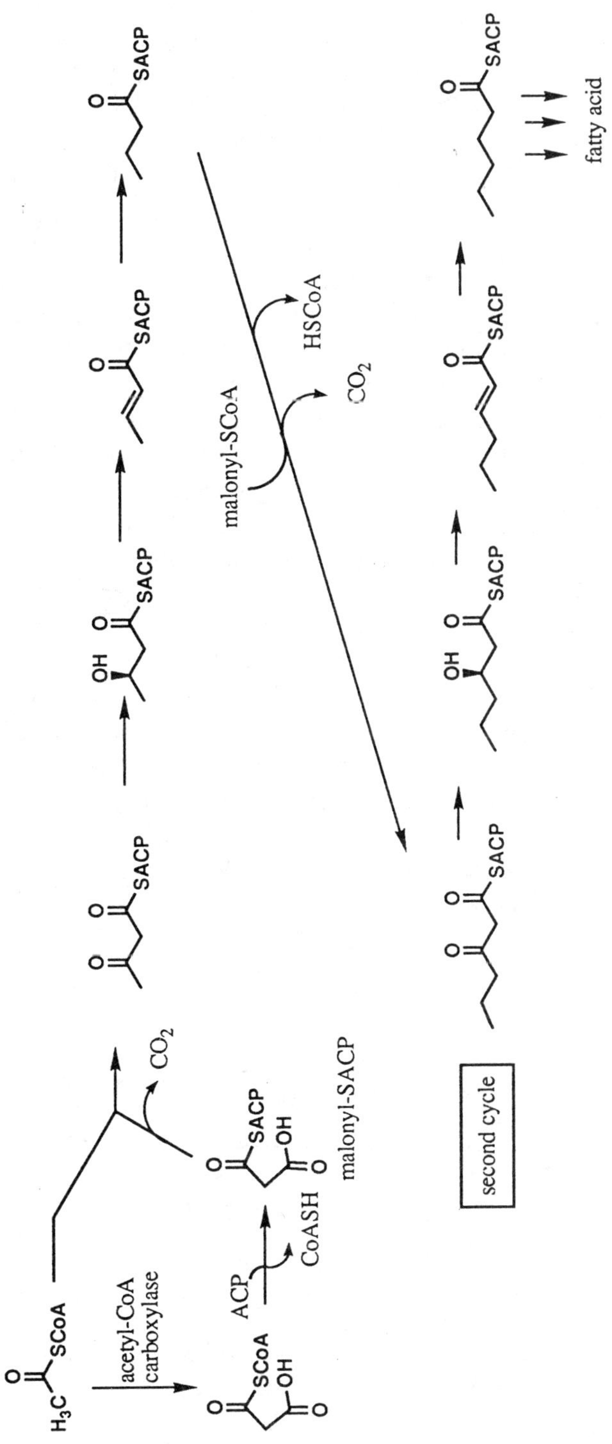

Fig. 1.1 — A summary of the first two cycles in fatty acid biosynthesis.

The second step consists of a series of partial reactions on acyl thioesters and is clearly the more complex. The key to the organizational control is an acyl carrier protein (ACP) (1.2) which chaperones the intermediate acyl thioesters around the catalytic sites of the FAS. In the dissociated Type II FASs this is a defined protein. In Type I FASs the catalytic domain of the multifunctional protein containing the serine residue which covalently links the 4′-phosphopantetheine group to the synthase is also termed the 'acyl carrier protein'. To initiate the second step malonyl-CoA undergoes a transacylation with the 4′-phosphopantetheine moiety of ACP and malonyl-ACP is relocated appropriately at the active site of the condensing enzyme, β-ketoacyl-ACP synthase. For the first condensation, a molecule of acetyl-CoA is then introduced into the active site of this enzyme such that a decarboxylative condensation between the two species can take place. The resultant β-ketoacyl-ACP is transported to the next enzyme, a β-ketoacyl-ACP reductase, where a reduction of the ketone, involving NADPH, affords D-β-hydroxyacyl-ACP. The ACP then guides the thioester through the final two reactions of the first cycle, i.e. dehydration by β-hydroxyacyl-ACP dehydratase, to the α,β-unsaturated thioester and then reduction by enoyl-ACP reductase to the saturated C_4-acyl-ACP (butyryl-ACP) intermediate. One cycle is now complete, and to continue the process the acyl moiety of butyryl-ACP is transfered to a free thiol residue at the active site of the condensing enzyme. The thiol of ACP is now released and is at liberty to accept another malonyl group by transesterification with malonyl-CoA, which it then orientates for condensation with the butyryl-SEnzyme to begin the second cycle. Condensation followed by the reduction sequence generates a C_6-acyl-ACP (hexanoyl-ACP). The process repeats itself, extending the acyl moiety by two carbon atoms after each cycle, until the optimum length of the FAS is reached and the fatty acid is released from the synthase complex. This is illustrated in Fig. 1.1

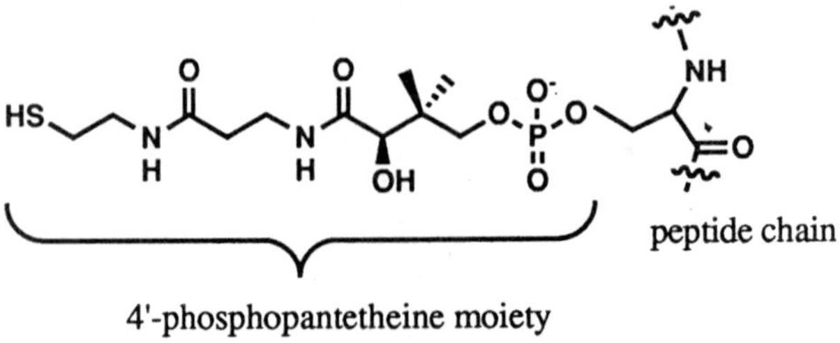

acyl carrier protein (1.2)

Until recently it was generally accepted that acetyl-CoA underwent a transacylation onto a free thiol residue of the condensing enzyme prior to the *first* condensation in fatty acid biosynthesis. A specific β-ketoacyl synthase (condensing enzyme III) has, however, been identified in *Escherichia coli* which mediates the first condensation between acetyl-CoA (not acetyl-ACP) and malonyl-ACP [3]. The β-ketoacyl-ACP produced is then processed by the enzymes of the FAS as discussed above. The direct condensation utilizing acetyl-CoA obviates the requirement for a transacylase activity loading acetyl-CoA onto a thiol residue of the condensing enzyme. This is proving to be a general phenomenon in fatty acid biosyntheses of the dissociated Type II FASs of bacteria and plants [4]. This condensing enzyme may be less discriminating towards the direct incorporation of coenzyme-A esters, a feature which may partially account for the variety of alternative starter units found to replace acetate in many fatty acids (see section 1.3.1).

1.1.1 Stereochemical course of fatty acid biosynthesis

The stereochemical course of all the relevant transformations of fatty acid biosynthesis has been established. In their classic investigations into these reactions Sedgwick and Cornforth utilized a combination of chiral methyl-acetyl-CoAs and malonyl-CoAs to explore the cryptic stereochemistry of these processes [5,6]. Although the methyl group is not chiral, chiral methyl acetyl-CoA can be generated by the stereospecific incorporation of hydrogen, deuterium and tritium isotopes on the methyl group. The chiral methyl group methodology is dependent on an intramolecular kinetic isotope effect which favours removal of hydrogen over deuterium over tritium. For instance, in the first reaction of fatty acid biosynthesis, acetyl-CoA carboxylase will preferentially break the C–H bond of chiral (2*S*)[2-^{2}H,2-^{3}H]-acetyl-CoA to generate malonyl-CoA which now has the diastereotopic C-2 hydrogen atoms differentially labelled. By tracing the fate of the tritium isotope, specific stereochemical information can be obtained. The results of Sedgwick and Cornforth's experiments are consistent with an overall *retention of configuration* for the carboxylation, giving rise to (2*S*)[2-^{2}H,2-^{3}H]-malonyl-CoA as shown in Fig. 1.2, i.e. the C–CO_2H bond of the product directly replaced the C–H bond of acetyl-CoA. Extension of this methodology also revealed that an *inversion of configuration* occurs during the course of the decarboxylative condensation mediated by the condensing enzyme. The action of β-ketoacyl reductase generates exclusively the (3*R*) or D-β-hydroxyacyl intermediate [7], and the subsequent dehydratase enzyme removes the elements of water by a *syn*-elimination of the D-3(3*R*)-alcohol and the 2(*pro*-2*S*)-hydrogen of the D-β-hydroxyacyl thioester [8].

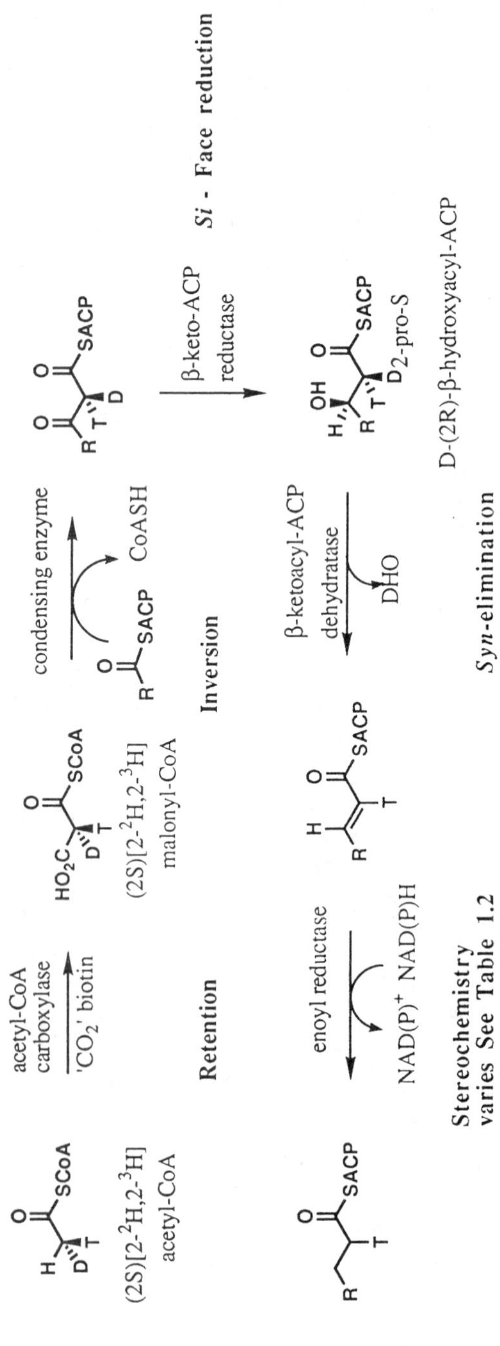

Fig. 1.2 — The stereochemical course of the reactions of fatty acid biosynthesis illustrated for (2S)[2-^{2}H,2-^{3}H] acetyl-CoA.

Up until this stage it would appear that the individual activities of FASs operate with a universal stereo-integrity; however, the stereochemical course of the final reducing enzyme, enoyl reductase, appears to be quite variable in different organisms. *trans*-2-Enoyl reductases from different sources can utilize different prochiral hydrogens of NAD(P)H and present hydrogens to varying faces of the double bond. The *E. coli* reductase for instance presents the *pro*-4*S* hydrogen of NADPH and adds hydrogens suprafacially to the 2*re*-3*si* face of the double bond. Yeast also utilizes the *pro*-4*S* hydrogen of NADPH but adds hydrogens antrafacially in a 2*si*-3*si* mode. The mammalian enzyme from rat liver FAS presents the *pro*-4*R* hydrogen of the co-factor to the *si* face with resultant incorporation into the *pro*-2*S* site of the saturated thioester. The available stereochemical data on this transformation from different sources is collated in Table 1.2. The stereochemical outcome at C-2 of a number of fungal strains (*A. cinerariae*

Table 1.2 — Stereochemistry and origin of various enoyl reductases

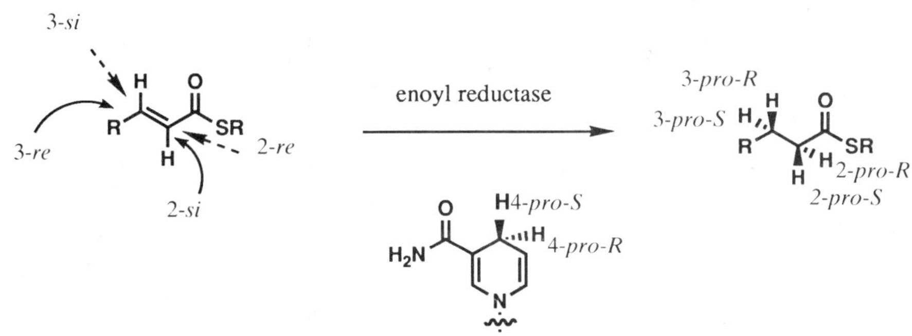

NAD(P)H

Source	NAD(P)H	C2	C3		Classification
E. coli	4-*pro-S*	*re*	*si*	P	bacteria
A. nidulands	—	*re*	—	P	blue-green algae
B. ammoniagenes	4-*pro-S*	*si*	*si*	E	bacteria
S. cerevisiae	4-*pro-S*	*si*	*si*	E	yeast
Rat liver	4-*pro-R*	*si*	*re*	E	mammal
P. brefeldianum	—	*si*	—	E	fungi
A. cinerariae	—	*si*	—	E	fungi
P. turbatum	—	*si*	—	E	fungi
C. cladosporioides	-	*si*		E	fungi
P. tricornutum	—	*re*		E	diatom
C. pyrenoidosa	—	*re*	—	E	green algae

(P) = prokaryote, (E) = eukaryote.

and *C. cladosporiodes*) has been investigated. These strains all produce polyketide secondary metabolites, and it is interesting that the enoyl reductase of the polyketide synthase presents hydride to the *re* face at C-2, whereas *in the same organism* the corresponding enzyme from the FAS presents hyride to the *si* face at C-2 (see section 3.4.4).

1.1.2 The production of triacetic acid lactone from fatty acid synthase

Purified preparations of FAS, when incubated with acetyl-CoA and malonyl-CoA, but in the absence of NADPH, generate triacetic acid lactone (TAL) (1.4) as a shunt or derailment product [9,10]. TAL appears to accumulate after cyclization of triacetic acid thioester (1.3), which itself results from condensation of malonyl-SR and acetoacetyl-ACP. These observations clearly indicate that ACP will escort the acetoacetyl moiety around the FAS and back to the condensing enzyme, even if the β-keto-functionality is not reduced. Presumably the FAS would continue to generate longer poly-β-keto chains in the absence of NADPH if the intermediates were not unstable and susceptible to cyclization.

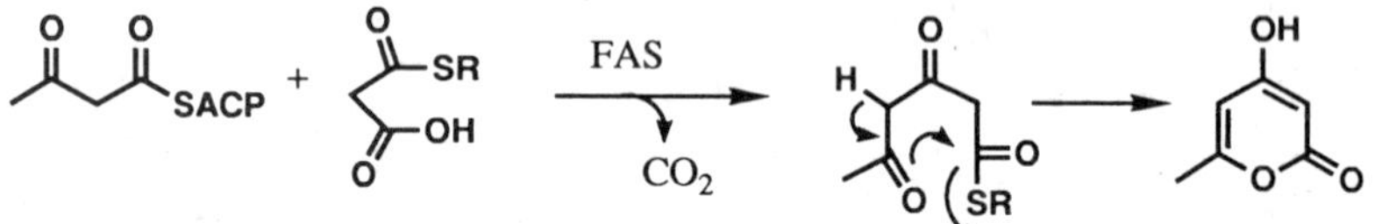

acetoacetyl-ACP malonyl-SR (1.3) triacetic acid lactone (TAL) (1.4)

TAL has also been detected in the absence of NADPH as a shunt metabolite in polyketide synthases of secondary metabolism [11] (see section 3.1.1) and it is tempting to speculate that those aromatic polyketides which derive from poly-β-keto-thioesters have their origin in FASs which have lost the ability to effect β-keto reduction but have developed a method for stabilizing the poly-β-keto intermediates.

1.2 TYPE I AND TYPE II FATTY ACID SYNTHASES

It is interesting at this stage to consider the different types of FASs which are expressed in various organisms. There appears to be a general increase in sophistication parallelling evolutionary complexity [12] (Table 1.3). For instance, most bacteria have a Type II FAS which consists of a *minimum* of seven discrete monofunctional enzymes and an ACP all of which are separable and express their individual activities in isolation from each other. The bacterial system studied in the greatest detail is palmitate synthase from *E. coli*, which appears to have two different types of enoyl reductase, one requiring NADPH and the second preferring NADH [13]. The first of these enzymes reduces the shorter chain lengths of C_4–C_6 while the second one is less specific and acts on the longer enoylacyl intermediates. *E. coli* also has three condensing enzymes. Condensing enzyme III mediates the first condensation with acetyl-CoA and malonyl-ACP, whereas condensing enzymes I and II mediate shorter and longer (C_{16} to C_{18}) chain extensions respectively.

Table 1.3 — FASs from different classes of organism

Organism	FAS constitution	*In vivo* structure	Type
Bacteria Plants Photosynthetic eukaryotes	At least seven discrete enzymes + acyl carrier protein	Dissociated–loosely associated	II
Yeast Fungi Some actinomycetes	Two multifunctional proteins	α_6, β_6 oligomer	I/II
Insects Birds Mammals	One multifunctional protein	Dimer for full activity	I

Type I FASs are found in most higher organisms except the plants. These are either tightly bound complexes of individual enzymes which do not dissociate easily, or more typically are enzyme complexes which contain multifunctional proteins. The more highly evolved mammalian FAS, for example, contains all the individual activities except the condensing activity on one peptide. *In vivo*, two almost identical multifunctional peptides come together to form a dimer, which expresses full catalytic activity. It would appear that dimerization is required to orientate the two free thiol groups of the condensing enzyme such that this final activity can be expressed [14].

Yeast and fungal FASs seem to have characteristics of both Type I and Type II complexes [15]. Yeast FAS consists fundamentally of two multifunctional proteins (α and β), each with complementary activities, the sum of which fully expresses saturated fatty acid biosynthesis. Both proteins must be associated, and in reality the molecular architecture is complex, with the synthase existing as an $\alpha_6\beta_6$ oligomer. The individual subunits are non-covalently bound and will dissociate in a manner characteristic of a Type II synthase. However, the multifunctional nature of the protein subunits represents a highly evolved system and places yeast FAS in a Type I category. It is noteworthy too that a few bacteria possess a Type I FAS and in particular it is representatives from the Actinomycetes which are emerging as the most significant in this regard [16]. Some Actinomycetes have FASs which appear to be clusters of multifunctional proteins with similar characteristics to the yeast and fungal systems. The Actinomycetes are particularly adept at generating complex polyketides as secondary metabolites, and the evidence from genetic analysis supports the involvement of multifunctional proteins in the biosynthesis of some secondary metabolites (see section 4.8.2).

The higher plants on the whole have a primitive bacterial Type II FAS associated exclusively with the chloroplasts. This is perhaps unexpected but is consistent with the endosymbiotic origin of chloroplasts from the photosynthetic cyanobacteria or blue-green algae [17]. The unicellular photosynthetic eukaryotic dinoflagellates are similar to the plants in this respect and have a Type II FAS associated with their chloroplasts, symptomatic of the prokaryotic origin of the chloroplast. *Euglena gracilis*, for example, has a Type II FAS located in the chloroplasts which is expressed only in the light, as well as a eukaryotic Type I FAS located in the cytoplasm [18].

1.3 SATURATED FATTY ACIDS

The common saturated fatty acids are even numbered and extend from C_8 to C_{20} as outlined in Table 1.1 [19]. Palmitic acid (C_{16}) is the most abundant fatty acid in Nature and tends to dominate over shorter chain fatty acids and stearic (C_{18}) acid in most bacteria, plants and animals. In plants, stearic acid is generated by a specific C_{16} elongase rather than a dedicated stearoyl synthase and longer chain fatty acids are subsequently elaborated in this stepwise manner. With few exceptions, animals only accumulate saturated fatty acids up to C_{20}. Of the marine organisms, sponges are significant as they can generate very long chain fatty acids (VLFAs) with lengths between C_{24} and C_{30} not uncommon [20]. Some bacteria biosynthesize fatty acids up to C_{26}, and in one dramatic instance the fermentation of *Mycobacterium smegmatis* (an Actinomycete) was manipulated such that 80% of the fats were C_{24} and C_{26}. In this case, and possibly with bacteria in general, it appears that palmitoyl synthase can generate these longer chain species when specific fermentation conditions are met [21].

1.3.1 Unusual saturated fatty acids

An interesting feature of many less common saturated fatty acids is a structural variation at the terminal end which arises as a consequence of a novel starter unit substituting for the usual acetyl-CoA moiety. The iso- (even and odd) and anteiso- series of fatty acids (Table 1.4) have starter units generated from isobutyryl-CoA, 2-methylbutyryl-CoA and 3-methylbutyryl-CoA, the catabolic products of the amino acids L-valine, L-leucine and L-isoleucine respectively. Some bacteria produce odd-numbered chain lengths, i.e. C_{15}, C_{17} and C_{19}, which derive from a propionyl-CoA starter unit, and, for example, *E. coli* will produce significant quantities of C_{15} and C_{17} fatty acids when incubated with propionate [22]. In a most interesting case, *Curtobacterium pusillum* produces ω-cyclohexylundecanoic acid (1.5) as its major lipid constituent (90%) under appropriate fermentation conditions, and the starter unit here is cyclohexylcarboxyl-SCoA, derived from shikimic acid [23]. *C. pusillum* is not unique in this respect and, for example, *Bacillus acidocaldrius*, a thermophilic bacterium, produces a range of saturated ω-cyclohexyl fatty acids [24,25].

Table 1.4 — Fatty acids with unusual starter units

Starter unit	Fatty acid	Series
Isobutyryl-CoA (from valine)	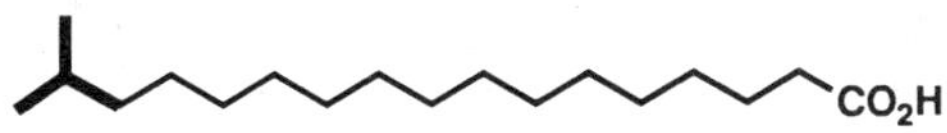	C_{18} iso-series (even)
2-Methylbutyryl-CoA (from leucine)	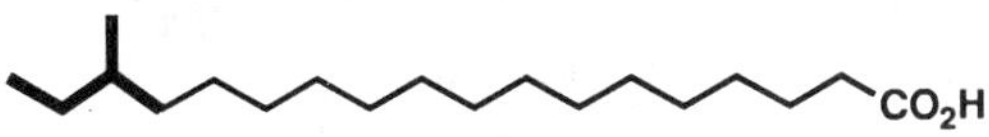	C_{19} iso-series (odd)
3-Methylbutyryl-CoA (from isoleucine)	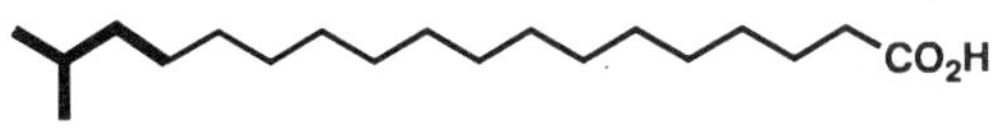	C_{19} anteiso-series (odd)
Fluoroacetyl-CoA	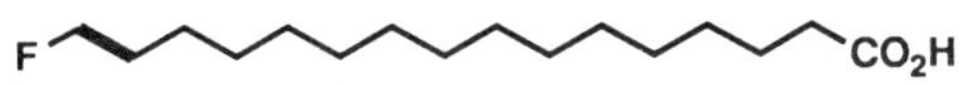	ω-Fluoropalmitic acid (1.6)
Propionyl-CoA		C_{17} (odd)
Cyclohexylcarboxyl-CoA	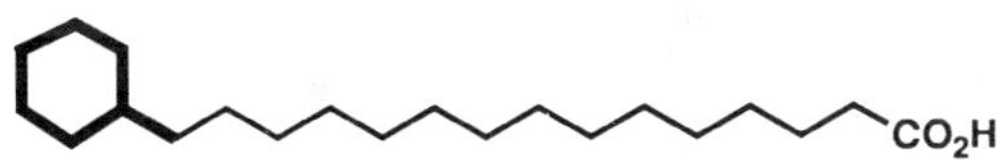	ω-Cyclohexyl-undecanoic acid (1.5)

An unusual and unique group of ω-fluorofatty acids including ω-fluoropalmitic acid (1.6) have been identified in seed oils of *Dichapetalum toxicarium*, a plant from Sierra Leone which accumulates the toxin fluoroacetate in its leaves [26]. These ω-fluorofatty acids are among a rare group of fluorinated natural products and they presumably arise from the incorporation of fluoroacetyl-CoA as a starter unit intercepting the normal biosynthetic machinery [27].

It is much less common to find fatty acids that will incorporate acetate surrogates *within* the chain, as this requires prior activation to the corresponding malonyl-CoA analogue. Multimethyl branched fatty acids derived from propionate (via methylmalonyl-CoA) are, however, natural products secreted from the uropygial gland of waterfowl, and it has been demonstrated, using cell-free preparations of this gland from geese, that propionate and methylmalonyl-CoA are incorporated into 2,4,6,8,-tetramethyldecanoic (1.7) and 2,4,6,8-tetramethylundecanoic (1.8) acids. Other fatty acids are also generated *in vitro* from the random incorporation of acetate and propionate subunits along the chain [28–30]. *In vitro* the FAS of the harderian gland of the guinea pig shows no discrimination between acetyl-CoA and propionyl-CoA as a starter unit and displays little selectivity for malonyl-CoA over methylmalonyl-CoA on chain extension [29]. The C_{32} fatty acid mycocerosic acid (1.9), from *Mycobacterium tuberculosis*, an Actinomycete bacteria responsible for tuberculosis, is a bizarre example, both in view of its length and incorporation of four terminal propionate subunits. Mycocerosic acid (1.9) is generated by methylmalonyl-CoA elongation of a C_{20} fatty acid [31]. Finally some insects such as termites and cockroaches, incorporate propionate via methylmalonyl-CoA into their surface lipids [32]. These observations are interesting within an evolutionary context as there are many more complex polyketide metabolites, particularly from Actinomycete bacteria, which incorporate propionate and even butyrate subunits (from ethylmalonyl-CoA) into their structural framework (see section 4.4.1).

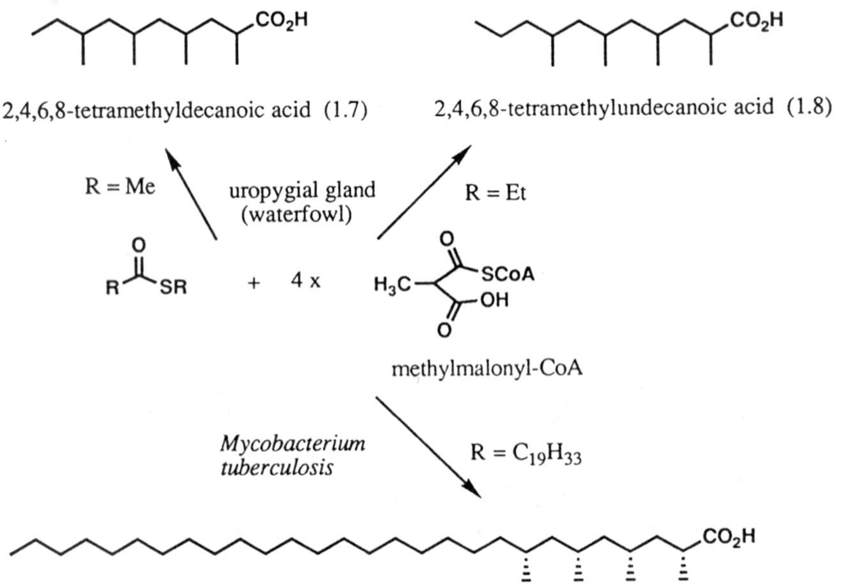

2R,4R,6R,8R-tetramethyloctacosanoic acid (mycocerosic acid) (1.9)

Cyclopropane and mid-chain methylated saturated fatty acids are found in some bacteria and to a lesser extent in plants and marine organisms. These compounds originate from *S*-adenosyl L-methionine (SAM) methylation of the double bond of an unsaturated fatty acid, usually bound as the acyl moiety of a phospholipid [33, 34]. The cyclopropane ring systems are derived from *cis*-double bonds and possess a *cis*-geometry. The most common examples are *cis*-9,10-methylenehexadecanoic acid (1.10) and *cis*-11,12-methyleneoctadecanoic (lactobacillic) acid (1.11) in bacteria, which originate from palmitoleate [16:1,(9*Z*)] and vaccenate [18:1,(11*Z*)] respectively.

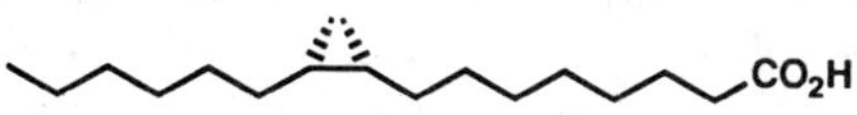

cis-9,10 - methylenehexadecanoic acid (1.10)

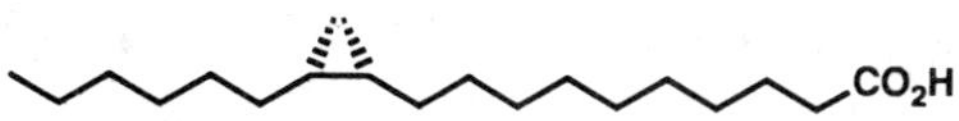

cis-11,12 - methyleneoctadecanoic acid (lactobacillic acid) (1.11)

Mid-chain methyl groups arise too from SAM, and the biosynthesis of 10-methylstearic acid (tuberculostearic acid) (1.14) from lipid-bound oleate in *Mycobacterium phlei* has been studied in some detail. Interestingly 10 *exo*-methylenestearate (1.13) was isolated and identified as an intermediate in the process, and it has been shown that the final step in the biosynthesis involves an NADPH-dependent reduction of the *exo*-methylene double bond [35].

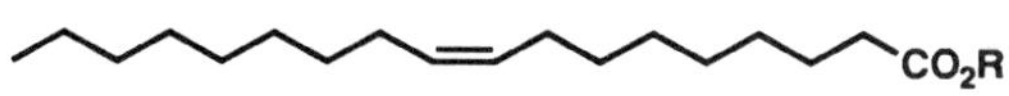

cis - oleate (1.12)

↓ (SAM)

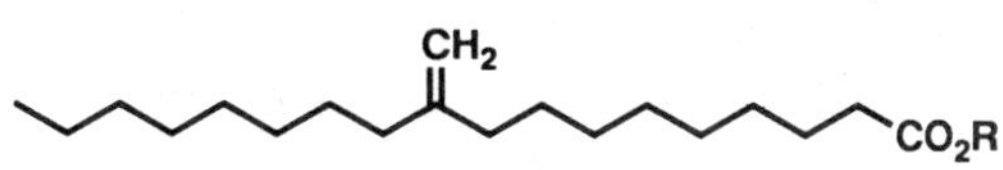

10 - *exo* - methylenestearate (1.13)

↓ NADPH

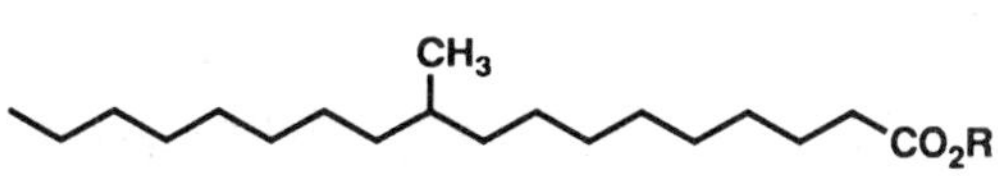

10 - methylstearate (1.14)

1.4 UNSATURATED FATTY ACIDS

Unsaturated fatty acids are utilized by organisms to control membrane fluidity. With few exceptions the double bonds are *cis* (or *Z*), introducing a bend into the extended zigzag hydrocarbon chain. The *cis*-geometry disrupts the efficient packing of the hydrophobic chains in fats and phospholipid membranes and increases their fluidity. A *trans* (or *E*) double bond induces little steric distortion with respect to saturated fatty acids in the packing of extended chains, as the zigzag conformation is retained, as illustrated in Fig. 1.3. Although *cis*-oleate (1.12) is widespread in all classes of organism, those which have evolved in colder environments tend to have increased levels of unsaturated (or branched chain) fatty acids in their lipids. Amazingly, ratios of unsaturated/saturated fatty acids can be manipulated by incubating many micro-organisms at various temperatures. For example, *Bacillus megaterium* grown at 35°C produces only saturated fatty acids; however, when the temperature is lowered to 20°C a desaturase enzyme is expressed and *cis*-unsaturated fatty acids are synthesized [36, 37].

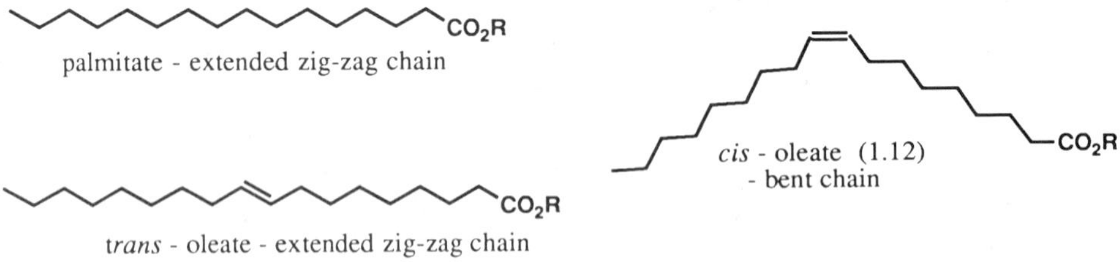

Fig. 1.3.

1.4.1 The introduction of double bonds into fatty acids

Double bonds are introduced into fatty acids by two different pathways, depending on the organisms involved. Bacteria (prokaryotes) synthesize mono-unsaturated fatty acids and have developed a process whereby the double bond is incorporated into the chain during assembly (anaerobic pathway). The higher eukaryotic organisms, on the other hand, modify the saturated hydrocarbon chain, *after* assembly, by a desaturation process that requires NAD(P)H and oxygen (aerobic pathway).

1.4.2 The anaerobic pathway

Bacteria all produce mono-unsaturated fatty acids and do so by chain extension of *cis*-3-decanoyl-ACP (1.17) a C_{10} acyl entity derived from D-β-hydroxydecanoyl-ACP (1.15), an intermediate in saturated fatty acid biosynthesis. The pivotal enzyme in this process is β-hydroxydeconyl-thioester dehydratase, which interconverts D-β-hydroxydeconyl-ACP and *cis*-3-decanoyl-ACP via a dehydration, to the intermediate *trans*-2-deconyl-SACP (1.16), followed by an allylic rearrangement. This enzyme switches the classical assembly of saturated fatty acids over to the synthesis of mono-unsaturated fatty acids. The stereochemical course of the allylic rearrangement between *trans*-2 (1.16) and *cis*-3-deconyl-ACP (1.17) has been elucidated and shown to be a 'suprafacial' process consistent with a strategically located base operating on one face of the double bond [38,39]. The 4(*pro*-4*R*) hydrogen is lost during the rearrangement and at C-2 a hydrogen atom is introduced to the *si* face to occupy the 2(*pro*-2*R*) site. Subsequent chain extension of *cis*-3-decanoyl-ACP (1.17) in the normal manner, generates palmitoleic (*cis*-9-hexadecenoic) (1.18) and vaccenic (*cis*-11-octadecenoic) (1.19) acids, the predominant unsaturated species in bacteria.

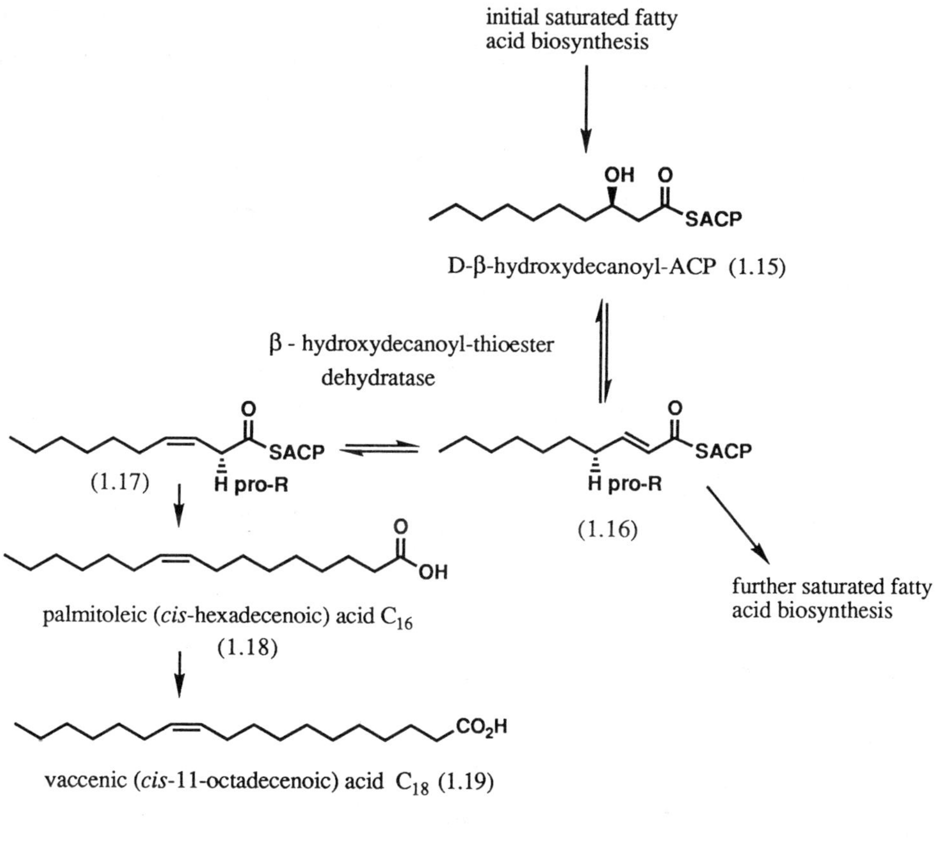

1.4.3 The aerobic (O_2-dependent) pathway

The aerobic or oxygen-dependent pathway is the more widespread process for the generation of unsaturated fatty acids in organisms. It is utilized by all eukaryote organisms and by some bacteria, and involves the modification of assembled saturated fatty acid thioesters (e.g. CoA thioesters in mammals and bacteria but ACP thioesters in plants).

$$RCH_2CH_2(CH_2)_nCOSR + NAD(P)H + H^+ + O_2 \rightarrow$$
$$RCH{=}CH(CH_2)_nCOSR + NAD(P)^+ + 2H_2O$$

The first desaturation usually occurs to introduce a *cis*-9,10 double bond, for example the generation of oleyl-CoA [18:1(9*Z*)](1.21) from stearyl-CoA, catalysed by stearyl-CoA desaturase. The substrate specificity of rat liver stearyl-CoA desaturase, for example, is not rigid, and it will accept a range of C_{12}–C_{19} saturated acyl-CoA thioesters; however the regioselectivity remains constant and the double bond is always introduced between C-9 and C-10 [40]. Two hydrogen atoms are stereospecifically abstracted from stearyl-CoA (1.20) during the process to generate the *cis* double bond, and it has been demonstrated in several systems that it is the *pro*-9*R* and *pro*-10*R* hydrogen atoms that are liberated [41,42].

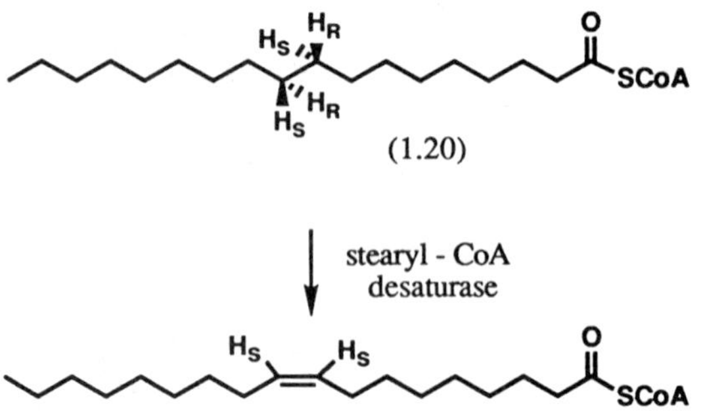

(1.20)

(1.21)

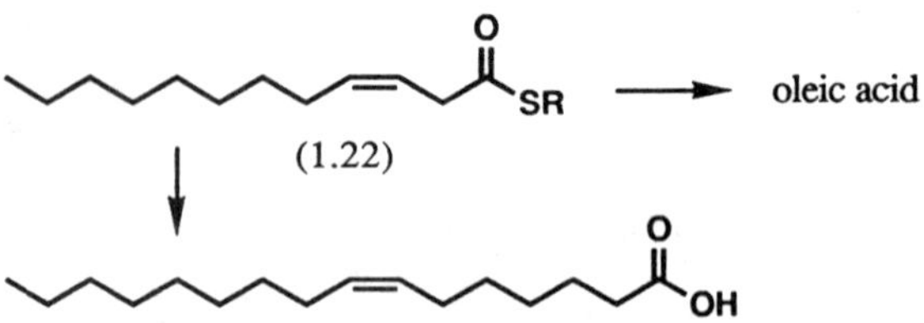

(1.22)

(1.23)

1.5 THE OCCURRENCE OF UNSATURATED FATTY ACIDS

1.5.1 Bacteria

The specificity of the anaerobic pathway limits the variety of unsaturated fatty acids found in bacteria, and palmitoleic (1.18) and *cis*-vaccenic (1.19) acids are the major unsaturated constituents of bacterial lipids. Some other mono-unsaturated fatty acids are found in individual micro-organisms. For example *Clostridium butyricum* produces *cis*-7-hexadecenoate (1.23) and oleic acids via the anaerobic route after chain extension from *cis*-3-dodecenoate (1.22)[43]. A few bacteria generate unsaturated fatty acids by employing the aerobic pathway. The substrates tend to be palmityl-CoA [16:0] and stearyl-CoA [18:0], which give rise to palmitoleic [16:1(9*Z*)] and oleic [18:1(9*Z*)] acids respectively. The aerobic pathway does not possess universal regiospecificity and certain bacteria, e.g. some *Bacillus* species, will generate mono-unsaturated fatty acids with double bonds at C-5, C-8, C-9 and C-10 [2]. With the exception of the cyanobacteria (blue-green algae), polyunsaturated fatty acids are unusual in prokaryotes and the bacterial polyenes which exist must be considered secondary metabolites unique to that organism.

1.5.2 Plants

The three major unsaturated fatty acid constituents in plants are oleic [18:1(9*Z*)], linoleic [18:2(9*Z*,12*Z*)] and linolenic [18:3(9*Z*,12*Z*,15*Z*)], which arise via the aerobic pathway. It is not uncommon to find lower levels of other unsaturated species, such as palmitoleic [16:1(9*Z*)] and the geometrically unusual *trans*-3-hexadecenoic acid [16:1(3*E*)](1.24). *cis*-Vaccenic acid [18:1(11*Z*)], typically found in bacteria, is relatively rare in plants, although some citrus fruits (mango, grapefruit, orange, etc.) can contain up to 20% in their lipid extracts [44]. The mono-unsaturated hydroxy acid ricinoleic acid (1.25) is not widely distributed but is important commercially as the major lipid constituent (90%) of castor oil and is derived via the oxidation of oleic acid.

Phospholipids of ferns, mosses and fungi are often abundant in arachidonic acid [20:4(5*Z*,8*Z*,11*Z*,14*Z*)](1.26), and for example the major lipid component (68–78%) of the fungus *Mortierella alpina* is composed of this acid [45]. Arachidonic acid (1.26) is a precursor of the prostaglandins in mammals and soft corals but has not been reported widely in higher plants. It is interesting that there are reports of prostaglandins occurring at low concentrations in flowering plants, and it may be that arachidonic acid is more widespread, in higher plants than is generally thought (see section 2.2.3).

The cyclopropene fatty acids malvalic (8,9-methylene-heptadec-8-enoic) (1.27) and sterculic (9,10-methylene-octadec-9-enoic) (1.28) acids are found in certain plant seed oils, and usually co-occur with their saturated cyclopropane analogues such as *cis*-9,10-methylenehexadecanoic acid (1.10) and lactobacillic acid (1.11) (see section 1.3.1), from which they are derived after desaturation [46].

Acetylenic fatty acids and their derivatives are secondary metabolites which are widespread and diverse in structure, not only in plants, but also in marine environments (see section 1.5.3). Acetylenic fatty acids have been identified widely in mosses and used as taxonomic markers [47]. Specific examples from plants include the C_{18} acetylene [18:2(6*A*,9*Z*)](1.29) produced by the liverwort *Riccia fluitans* [48], and crepenynic (*cis*-9-octadecene-12-ynoic) (1.30) isolated from the higher Compositae (daisy). A diverse range of acetylenic thiophenes such as (1.31) are widely distributed among flowering plants but in particular among the Heliantheae, a sub group of the Compositae, and these are believed to derive after elaboration of acetylenic fatty acids [49].

(1.24)

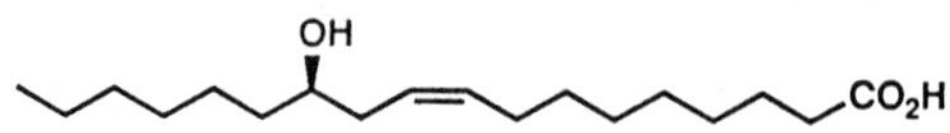

(1.25)

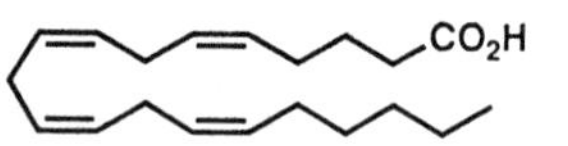

(1.26)

(1.27)

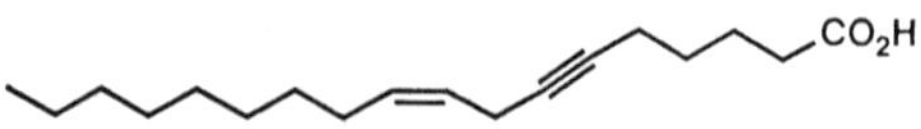

(1.29)

(1.28)

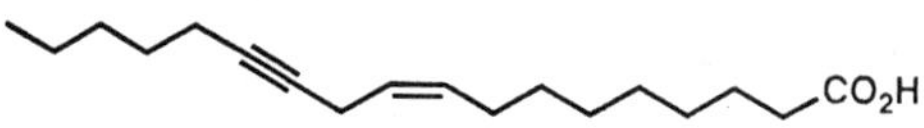

(1.30)

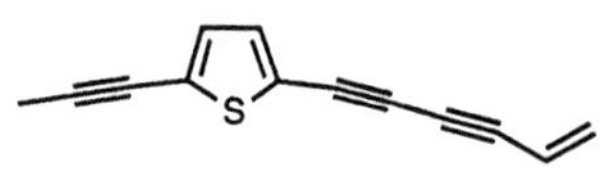

(1.31)

1.5.3 Marine organisms

Marine organisms extend from the cyanobacteria (prokaryotes) and eukaryotic plankton through the algae to the invertebrates such as mollusks and sponges, and collectively they elaborate a great variety of unsaturated fatty acid structures. Oleic acid [18:1(9*Z*)] is common throughout as a primary metabolite, and some cyanobacteria synthesize linoleic [18:2(9*Z*,12*Z*)] and linolenic [18:3(9*Z*,12*Z*,15*Z*)] acids; however, the variety becomes apparent when the secondary metabolites are considered (see also Chapter 5). Polyunsaturated compounds containing up to five double bonds are not unusual. *Ptilota filicina* (red algae), for example, produces the polyunsaturated fatty acids [20:5(5*E*,7*E*,9*E*,15*Z*,18*Z*)] (1.32) and [20:5(5*Z*,7*E*,9*E*,15*Z*,18*Z*)] (1.33) [50]. Related hydroxylated analogues have been identified from the British coastal red algal species *Laurencia hybrida* [51].

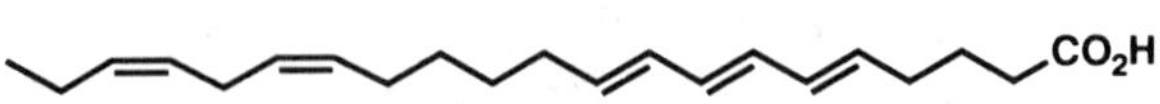

(1.32)

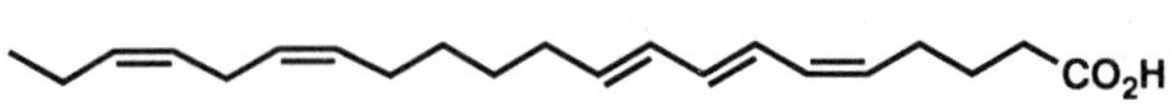

(1.33)

Sponges generate an interesting and diverse array of functionalized fatty acids as lipid constituents, and in particular they are known to elaborate very long chain fatty acids (VLFAs) (C_{24}–C_{34}), including straight chain and branched unsaturated species. The fatty acids (1.34) and (1.35) are typical and originate from chain extension of shorter saturated fatty acids, i.e. palmitic acid is a precursor of (1.34) whereas 13-methyltetradecanoic acid (iso-15:0) (1.36) is a precursor of (1.35). The double bonds of the great majority of VLFAs isolated from sponges predominate at the C-5 and C-9 sites, indicating a regiospecific desaturase system operating after chain elongation [52]. Many more complex fatty acids have been isolated from sponges and the brominated acetylenes (1.37) and (1.38) serve to illustrate how elaborate some of these structures are [53]. Interestingly a series of prostaglandins (derivatives of C_{20} polyunsaturated fatty acids) have been isolated from soft corals and these are discussed in sections 2.2.1 and 2.3.2.

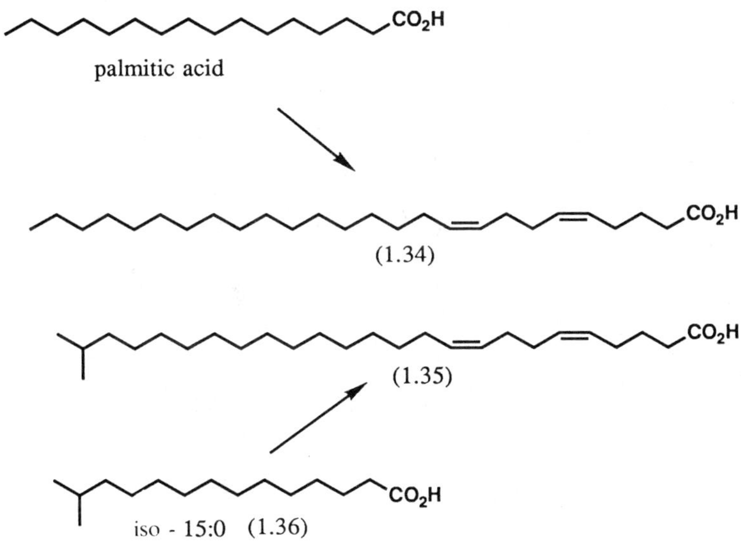

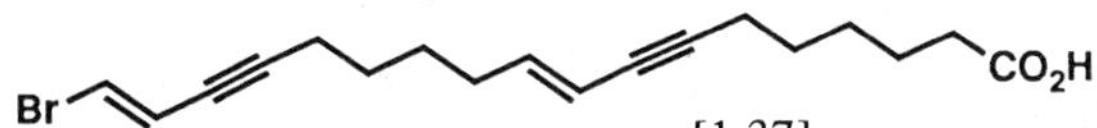

[1.37]

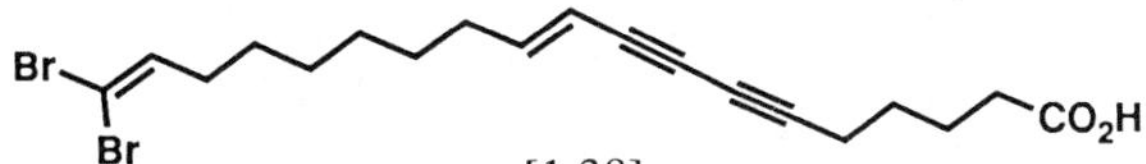

[1.38]

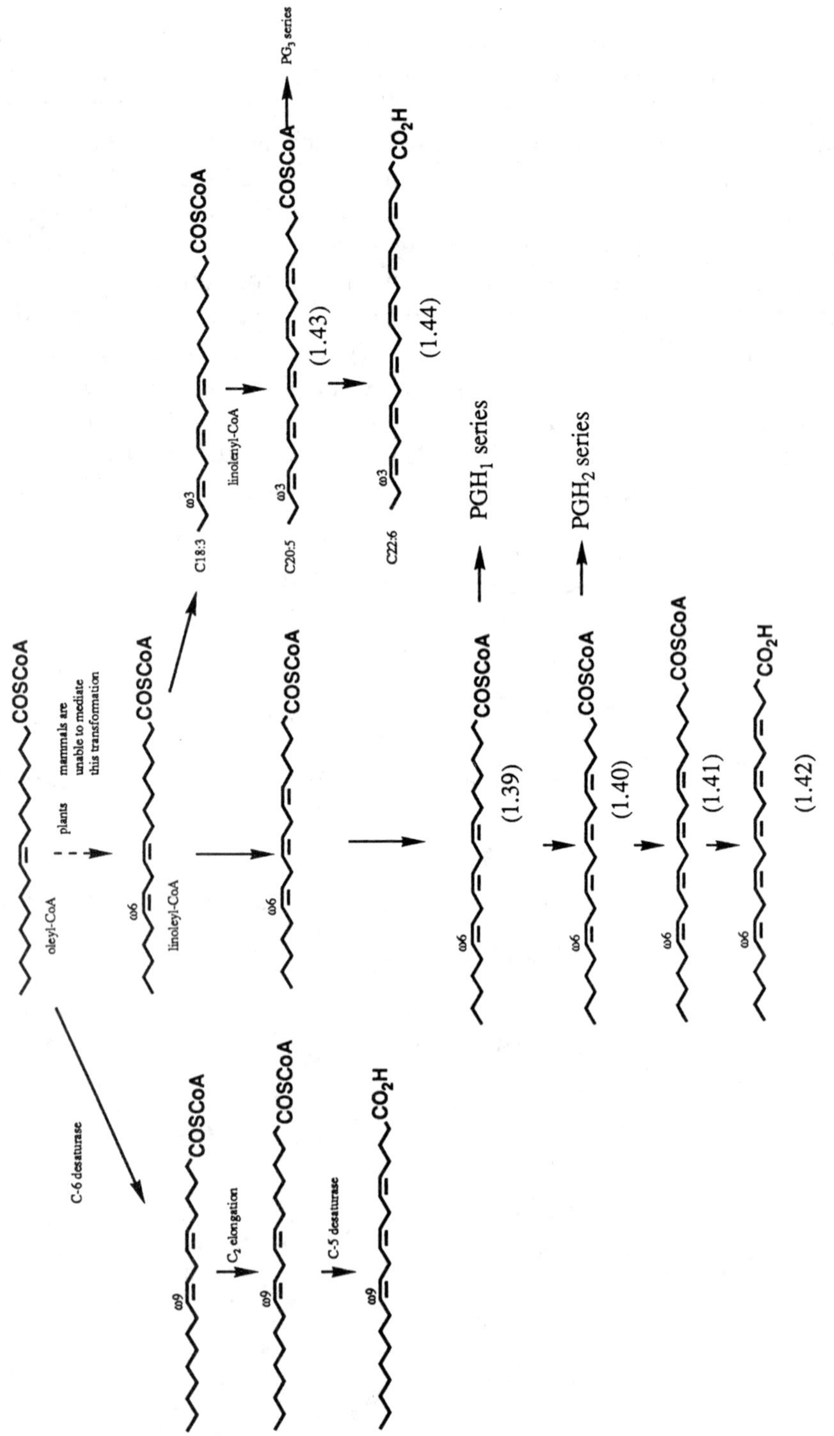
oleyl-CoA
COSCoA
plants
mammals are unable to mediate this transformation
C-6 desaturase
ω6
linoleyl-CoA
C18:3
ω3
linolenyl-CoA
C20:5
C22:6
PG$_3$ series
CO$_2$H
(1.43)
(1.44)
PGH$_1$ series
PGH$_2$ series
(1.39)
(1.40)
(1.41)
(1.42)
ω9
C$_2$ elongation
C-5 desaturase

Fig. 1.4.

1.5.4 Mammalian eicosaenoic acids

The ω3, ω6 and ω9 eicosaenoic fatty acids represent the three groups of polyunsaturated fatty acids found in mammals. The ω9 pathway is distinct from the ω3 and ω6 pathways and arises directly from metabolism of oleoyl-CoA.

A significant feature of mammalian unsaturated fatty acid metabolism is the inability of animals to convert oleyl-CoA into linoleyl-CoA, as mammals are devoid of a C-12 desaturase. Linoleic acid is therefore an *essential* fatty acid (dietary requirement) obtained from plants and is the precursor to the important ω3 and ω6 groups.

The ω3 and ω6 pathways ultimately generate highly unsaturated C_{22} compounds. On the ω6 pathway [20:3(8*Z*,11*Z*,14*Z*)] (1.39) and particularly arachidonyl-CoA [20:4,(5*Z*,8*Z*,11*Z*,14*Z*)] (1.40) are significant as precursors to the prostaglandins (PGH_1 and PGH_2 series respectively) and related hormones (see Chapter 2). Arachidonyl-CoA (1.40) can also undergo further chain extension to [22:4(7*Z*,10*Z*,13*Z*,16*Z*)] (1.41) and then desaturation to [22:5(4*Z*,7*Z*,10*Z*,13*Z*,16*Z*)] (1.42). Similar processes are operating on the ω3 pathway which lead to the hexaunsaturated fatty acid [22:6(4*Z*,7*Z*,10*Z*,13*Z*,16*Z*,19*Z*)] (1.44). The pentaunsaturated intermediate [20:5(5*Z*,8*Z*,11*Z*,14*Z*,17*Z*)] (1.43) on this pathway is important as a precursor to the PGH_3 prostaglandin series (see section 2.3.1). The main pathways and intermediates in mammalian unsaturated fatty acid biosynthesis are illustrated in Fig. 1.4.

REFERENCES

[1] S. J. Wakil, J. K. Stoops and V. C. Joshi, Fatty acid synthesis and its regulation. *Ann. Rev. Biochem.*, 1983, **52**, 537.

[2] A. J. Fulco, Fatty acid metabolism in bacteria. *Prog. Lipid Res.*, 1983, **22**, 133.

[3] S. Jackowski, C. M. Murphy, J. E. Cronan and C. O. Rock, Acetoacetyl-acyl carrier protein synthase: a target for the antibiotic thiolactomycin. *J. Biol. Chem.*, 1989, **264**, 7624.

[4] M. C. Walsh, W. E. Klopfenstein and J. L. Harwood, The shortchain condensing enzyme has a widespread occurrence in the fatty acid synthetases from higher plants. *Phytochemistry*, 1990, **29**, 3797.

[5] B. Sedgwick and J. W. Cornforth, The biosynthesis of long-chain fatty acids. Stereochemical differentiation and enzyme incorporation of chiral acetates. *Eur. J. Biochem.*, 1977, **75**, 465.

[6] B. Sedgwick, J. W. Cornforth, S. J. French, R. T. Gray, E. Kelstrup and P. Willadsen, The biosynthesis of long-chain fatty acids. Incorporation of radioactivity from stereospecifically tritiated malonyl thiolesters

and stereochemistry of the acetyl-CoA carboxylase reaction. *Eur. J. Biochem.*, 1977, **75**, 481.
[7] S. J. Wakil and R. Bresser, Studies on the mechanism of fatty acid synthesis. *J. Biol. Chem.*, 1962, **237**, 687.
[8] B. Sedgwick, C. Morris, and S. J. French, Stereochemical course of dehydration catalysed by the yeast fatty acid synthetase. *J. Chem. Soc. Chem. Commun.*, 1978, 193.
[9] J. E. Nixon, J. R. Putz and J. W. Porter, Synthesis of triacetic acid lactone by the pigeon liver fatty acid synthetase complex. *J. Biol. Chem.*, 1968, **243**, 5471.
[10] M. Yalpani, K. Willecke and F. Lynen, Triacetic acid lactone, a derailment product of fatty acid biosynthesis. *Eur. J. Biochem.*, 1969, **8**, 495.
[11] A. I. Scott, L. C. Beadling, N. H. Georgopapadakou and C. R. Subbarayan, Biosynthesis of polyketides. Purification and inhibition studies of 6-methylsalicylic acid synthase. *Bioorganic Chem.*, 1974, **3**, 238.
[12] A. D. McCarthy and D. G. Hardie, Fatty acid synthase, an example of protein evolution by gene fusion. *TIBS.*, 1984, 60.
[13] J. J. Volpe and P. R. Vagelos, Saturated fatty acid biosynthesis and its regulation. *Ann. Rev. Biochem.*, 1973, **42**, 21.
[14] J. S. Buckner and P. E. Kolattukudy, One step purification and properties of a two peptide fatty acid synthetase from the uropygial gland of the goose. *Biochemistry*, 1976, **15**, 1948.
[15] F. Weiland, L. Renner, C. Verfurth and F. Lynen, Studies on the multienzyme complex of yeast fatty acid synthetase. *Eur. J. Biochem.*, 1979, **94**, 189.
[16] H. W. Knoche and K. E. Koths, Characterisation of a fatty acid synthetase from *Cornebacterium diphtheriae*. *J. Biol. Chem.*, 1973, **248**, 3517.
[17] P. K. Stumpf. The biosynthesis of saturated fatty acids, *The Biochemistry of Plants. A Comprehensive Treatise*. Vol 9, Chapter 5. Eds P. K. Stumpf and E. E. Conn, 1987, Academic Press.
[18] R. W. Hendren and K. Bloch, Fatty acid synthetases from *Euglena gracilis*. *J. Biol. Chem.*, 1980, **255**, 1504.
[19] T. P. Hilditch and P. N. Williams, The chemical construction of the natural fats. 1974, 4th Edition, Wiley.
[20] P. R. Bergquist, M. P. Lawson, A. Lavis and R. C. Cambie, Fatty acid composition and the classification of the *Porifera*. *Biochem., Syst., Ecol.*, 1984, **12**, 63.
[21] K. Bloch and D. Vance, Control mechanisms in the synthesis of saturated fatty acids. *Ann. Rev. Biochem.*, 1977, **46**, 263.

[22] L. O. Ingram, L. S. Chevalier, E. J. Gabbay, K. D. Ley and K. J. Winters, Propionate induced synthesis of odd chain length fatty acids by *Escherichia coli*. *J. Bact.*, 1977, **131**, 1023.

[23] K. I. Suzuki, K. Saito, A. K. Kawaguchi, S. Okuda and K. Komagata, Occurrence of ω-cyclohexyl fatty acids in *Curtobacterium pusillum*. *J. Gen. Appl. Microbiol.*, 1981, **27**, 261.

[24] M. DeRosa, A. Gambacorta, L. Minale and J. D. Bu'lock, Cyclohexane fatty acids from a thermophilic bacterium. *J. Chem. Soc. Chem. Commun.*, 1971, 1334.

[25] A. Blume, R. Dreber and K. Prolla, The influence of branched chain ω-alicyclic fatty acids on the transition temperature of *Bacillus subtilis* lipids. *Biochem. Biophys. Acta*, 1978, **512**, 489.

[26] P. F. V. Wood, R. J. Hall and R. A. Peters, Fluorofatty acids in seeds of *Dichapetalum toxicarium*. *Nature*, 1964, **201**, 611.

[27] D. B. Harper, J. T. G. Hamilton and D. O'Hagan, Identification of *threo*-18-fluoro-9,10-dihydroxystearic acid: a novel ω-fluorinated fatty acid from *Dichapetalum toxicarium* seeds. *Tetrahedron Letts.*, 1990, 7661.

[28] J. S. Buckner and P. E. Kolattukudy, Lipid biosynthesis in sebaceous glands: regulation of the synthesis of *n*- and branched chain fatty acids by malonyl-coenzyme-A decarboxylase. *Biochemistry*, 1975, **14**, 1768.

[29] Y. Seyama, H. Otsuka, A. Kawaguchi and T., Yamakawa, Fatty acid synthetase from the harderian gland of guinea pig: biosynthesis of methyl branched fatty acids. *J. Biochem.*, 1981, **90**, 789.

[30] J. S. Buckner, P. E. Kolattukudy and L. Rogers, Synthesis of multimethyl branched fatty acids by avian and mammalian fatty acid synthetase and its regulation by malonyl-CoA decarboxylase in the uropygial gland. *Arch. Biochem. Biophys.*, 1978, **186**, 152.

[31] D. L. Rainwater and P. E. Kolattukudy, Synthesis of mycocerosic acids from methylmalonyl-coenzyme-A by cell free extracts of *Mycobacterium tuberculosis* var *bovis* BCG. *J. Biol. Chem.*, 1983, **258**, 2979.

[32] J. Chase, R. A. Jurenka, C. Schal, P. P. Halarnkar and G. Blomquist, Biosynthesis of methyl branched hydrocarbons of the German cockroach, *Blattella germanica* (L.) (Orthoptera, Blattellidae). *Insect Biochem.*, 1990, **20**, 149.

[33] H. Zalkin, J. H. Law and H. J. Goldfine, Enzymatic synthesis of cyclopropane fatty acids catalysed by bacterial extracts. *J. Biol. Chem.*, 1963, **238**, 1242.

[34] P. H. Buist and D. B., MacLean, The biosynthesis of cyclopropane fatty acids II. Mechanistic studies using methionine labelled with one, two and three deuterium atoms in the methyl group. *Can. J. Chem.*, 1982, **60**, 371.

[35] Y. Akamatsu and J. H. Law, Enzymatic alkylenation of phospholipid fatty acid chains by extracts of *Mycobacterium phlei*. *J. Biol. Chem.*, 1970, **245**, 701.

[36] A. J. Fulco, The biosynthesis of unsaturated fatty acids by *Bacilli* II. Temperature dependent biosynthesis of polyunsaturated fatty acids. *J. Biol. Chem.*, 1970, **245**, 2985.

[37] D. Fujii and A. J. Fulco, Biosynthesis of unsaturated fatty acids by *Bacilli*. Hyperinduction and modulation of desaturase synthesis. *J. Biol. Chem.*, 1977, **252**, 3660.

[38] J. M. Schwab and J. B. Klassen, β-Hydroxydeconyl-thioester dehydratase. Steric course at substrate C-4 in the enzyme catalysed allylic rearrangement. *J. Chem. Soc. Chem. Commun.*, 1984, 296.

[39] J. M. Schwab and J. B. Klassen, β-Hydroxy-thioester dehydratase. Steric course at substrate C-2 and overall steric course of the enzyme catalysed allylic rearrangement. *J. Chem. Soc. Chem. Commun.*, 1984, 298.

[40] H. G. Enoch, A. Catala and P. Strittmater, Mechanism of rat liver microsomal stearyl-CoA desaturase. *J. Biol. Chem.*, 1976, **251**, 5095.

[41] G. J. Schroepfer Jr and K. Bloch, The stereospecific conversion of stearic acid to oleic acid. *J. Biol. Chem.*, 1965, **240**, 54.

[42] L. J. Morris, Mechanisms and stereochemistry in fatty acid metabolism. *Biochem. J.*, 1970, **118**, 681.

[43] G. Scheuerbrandt, H. Goldfine, E. Baronowsky and K. Bloch, A novel mechanism for the biosynthesis of unsaturated fatty acids. *J. Biol. Chem.*, 1961, **236**, PC 70.

[44] A. Shibahara, K. Tamamoto, T. Nakayama and G. Kajimoto, *cis*-Vaccenic acid in pulp lipids of commonly available fruits. *J. Am. Oil. Chem. Soc.*, 1987, **64**, 397.

[45] N. Totani and K. Oba, The filamentous fungus *Mortierella alpina*, high in arachidonic acid. *Lipids*, 1987, **22**, 1060.

[46] E. M. Gaydou and A. R. P. Ramanoelina, A survey of the *Sarcolaenaceae* for cyclopropene fatty acids. *Phytochemistry*, 1983, **22**, 1725.

[47] G. Kohn, S. Demmerke, O. Vandekerkhove, E. Hartman and P. Beutelmann, Distribution and chemotaxonomic significance of acetylenic fatty acids in mosses of *Dicranale*. *Phytochemistry*, 1987, **26**, 2271.

[48] G. Kohn, A. Vierengel, O. Vandekerkhove and E. Hartman, 9-Octadecene-6-ynoic acid from *Ricca fluitans*. *Phytochemistry*, 1987, **26**, 2101.

[49] L. P. Christensen and J. Lam, Acetylenes and related compounds in Heliantheae. *Phytochemistry*, 1991, **30**, 11.

[50] A. Lopez and W. H. Gerwich, Two new icosapentaenoic acids from the temperate red seaweed *Ptilota filicina* J. Arardh. *Lipids*, 1987, **22**, 1980.

[51] M. D. Higgs, Antimicrobial components of the red alga *Laurencia hybrida* (Rhodophyta, Rhodomelaceae). *Tetrahedron, 1981,* **37**, 4255.

[52] C. Djerassi and W.-K. Lam, Sponge phospholipids. *Acc. Chem. Res., 1991,* **24**, 69.

[53] S. Hirsh, S. Carmley and Y. Kashman, Brominated unsaturated acids from the marine sponge *Xestospongia sp. Tetrahedron, 1987,* **43**, 3257.

2

Prostaglandins and related hormones

In the early 1930s it became apparent that there existed biologically active constituents in the lipid fraction of human semen [1–4]. The induced responses varied and ranged from the contraction and relaxation of uterine and intestinal smooth muscle to the lowering of blood pressure. Ulf von Euler introduced the term 'prostaglandin' at this time as he believed (erroneously) that the active compounds were secreted by the prostate gland and accumulated in the seminal vesicles [4]. The term has found general acceptance in spite of von Euler suggesting the more appropriate term 'vesiglandin' in a paper in 1936 [5]. Since these early observations a great deal of effort has gone into the isolation and structural elucidation of the prostanoid hormones as well as unravelling details of their biosynthesis.

In general the prostaglandins are not transported in the blood but are short-lived 'local' hormones which act at their site of synthesis and are then rapidly metabolized. The prostanoid hormones are present and active at low concentrations and coupled, with their metabolic instability, it is a significant achievement that these compounds have been isolated and their structures determined. Although prostanoids are produced by most organs of the body, they are secreted in the greatest abundance by the seminal vesicles of humans and sheep and are found in reasonable quantities in the semen. Curiously the horse, ox, rabbit and dog are devoid of prostaglandins from this source [6].

2.1 PROSTAGLANDIN STRUCTURE

The prostaglandins are elaborated C_{20} unsaturated fatty acids, containing a cyclopentane ring system, and are grouped into three structural series. In general, all prostaglandins have a *trans* double bond at C-13 and a (15*S*) hydroxyl group. Those with only this double bond belong to Series 1.

Members of Series 2 have an additional *cis* double bond at C-5 and those in Series 3 have a third *cis* double bond at C-17. Further classification focuses on the functionality of the ring system and the group designation A, B, C, D, E, F_α and F_β is applied to the partial structures shown in Fig. 2.1.

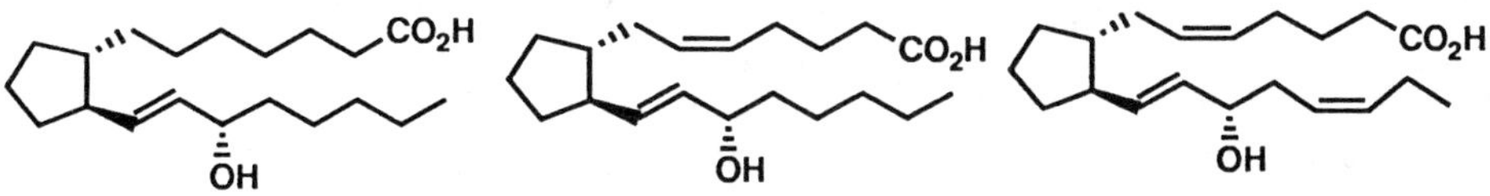

PG_1 PG_2 PG_3

basic structural features of PG_1, PG_2 and PG_3.

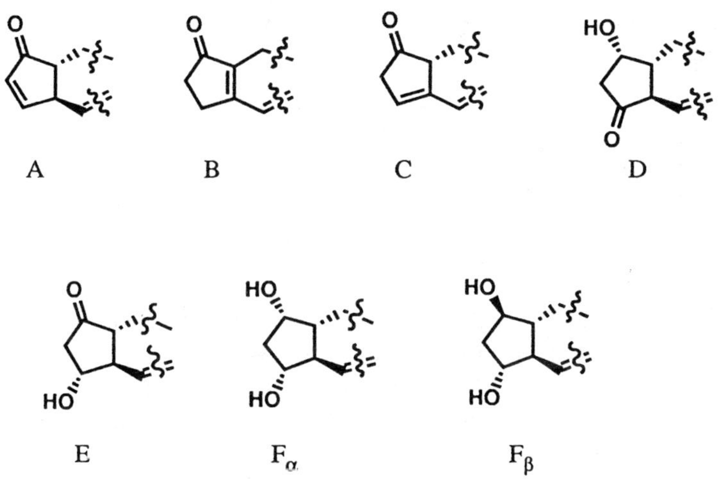

Fig. 2.1 — The structures of the prostaglandins.

It took almost thirty years from their discovery before the structures and stereochemistry of the first prostaglandins were reported in the early 1960s by Bergstrom and his co-workers [7,8]. They isolated and identified the six primary compounds of the E and F groups, PGE_1, PGE_2, PGE_3, $PGF_{1\alpha}$, $PGF_{2\alpha}$ and $PGF_{3\alpha}$ from sheep vesicular glands. Subsequently three further biologically active constituents were isolated and their structures determined as the three serial members of the D group, PGD_1, PGD_2 and PGD_3 [9,10]. Presently there appear to be nine primary prostaglandins secreted in mammals and these compounds are illustrated in Fig. 2.2.

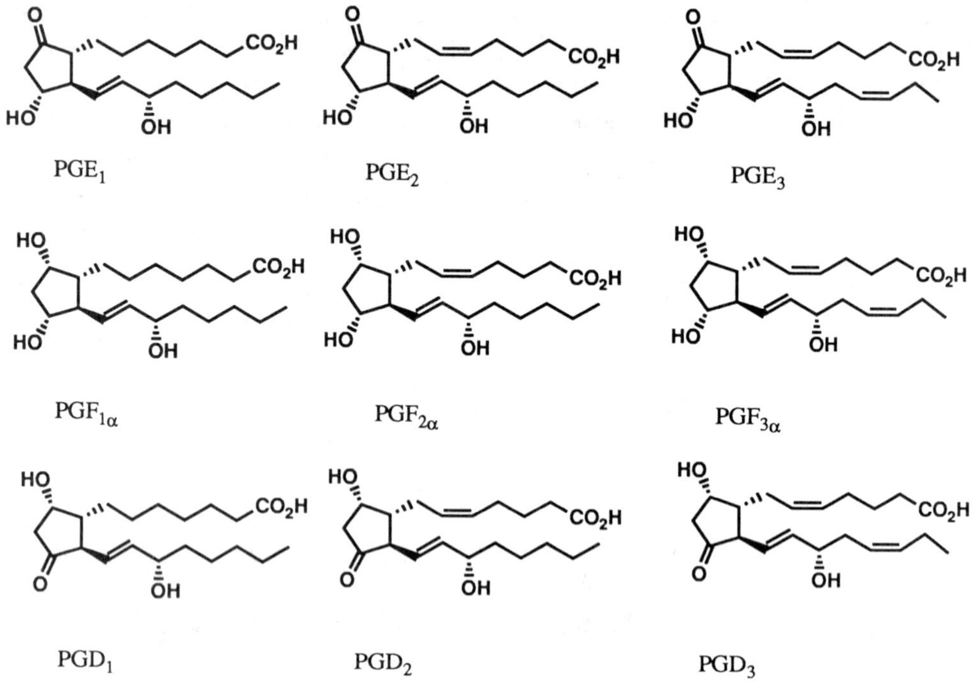

Fig. 2.2 — The nine primary mammalian prostaglandins.

2.2 THE NON-MAMMALIAN OCCURRENCE OF PROSTAGLANDINS

Clearly the discovery of the prostaglandins and the sustained interest and interrogation of these hormones stems from their mammalian origin and the physiological responses they induce, particularly in humans. The prostaglandins, however, are not exclusively mammalian natural products and indeed have been identified from many different sources. Characteristically prostaglandins do not accumulate in tissue and tend to have a transient existence; predictably therefore, their detection is difficult and the extent of their occurrence is only emerging.

2.2.1 Marine corals

Perhaps surprisingly, marine soft corals have proven to be a rich source (and resource) of prostanoid compounds. The most significant example of this is the production of the (15*R*)-acetoxy methyl ester (2.1) of PGA_2 and its hydrolysed products (1.5% dry weight) by *Plexaura homomalla* [11]. The mammalian prostaglandins possess the (15*S*) stereochemistry, and variants of *P. homomalla* have been found [12] which secrete the epimeric (15*S*)-15-acetoxy methyl ester (2.2) of PGA_2 (1.5–2.0% dry weight). This variant also co-produces smaller amounts of the 5-*trans*-isomer (2.3) of prostaglandin PGA_2.

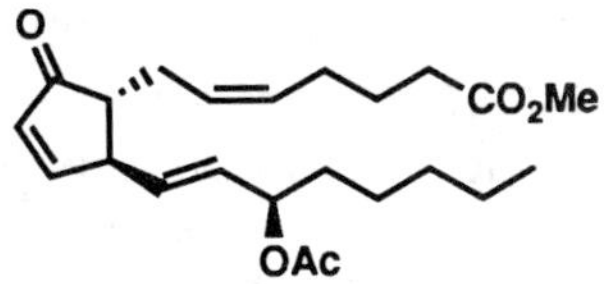

(2.1)

from *Plexaura homomalla*

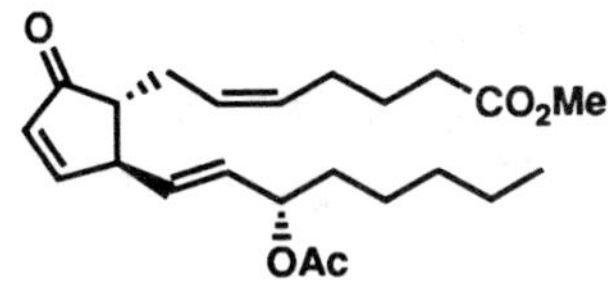

(2.2)

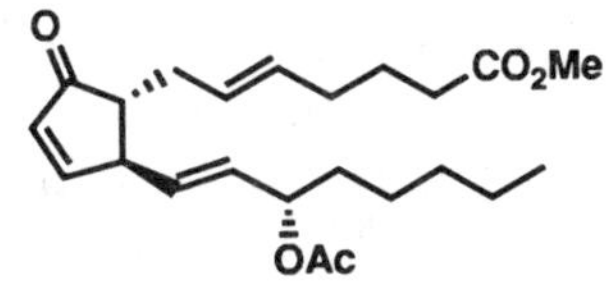

(2.3)

The soft coral *Clavularia viridis* also secretes a series of prostanoid compounds, which have been isolated independently by two research groups and termed the claviridenones [13,14] or clavulones-A to -D (2.4–2.9) [15]. Again these are acetoxy methyl esters of the PGA group. $PGF_{2\alpha}$ has been isolated from the Japanese gorgonean *Euplexaura erecta* [16], and $PGF_{2\alpha}$ 11-acetate methyl ester and its 18-acetoxy analogue have been identified in the Red Sea soft coral *Lobophyton depressum* [17].

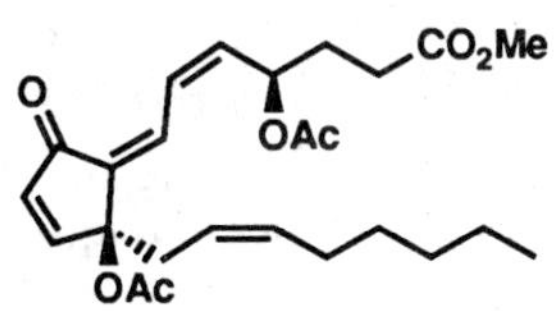

(2.4)

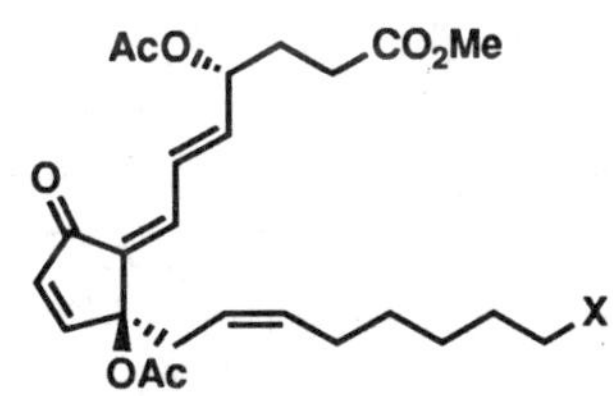

X = H (2.5)
X = OAc (2.6)

from *Clavularia viridis*

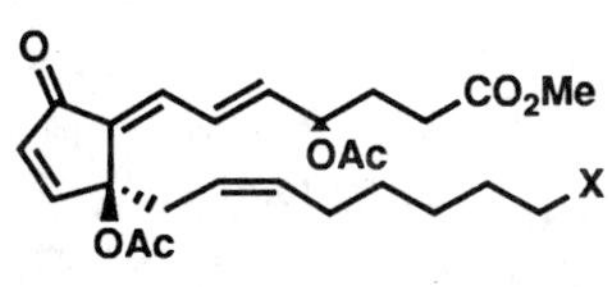

X = H (2.7)

X = OAc (2.8)

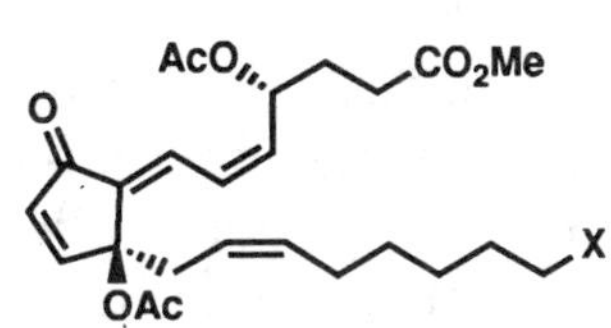

(2.9)

The high concentrations of these compounds in coral tissue is anomalous and suggests that they are not involved in any physiological control mechanisms of the host organisms. They may clearly have a role to play in defence against predators. The biosynthesis of the mammalian and marine prostaglandins is quite distinct (see section 2.3.2) and it is interesting that selective evolutionary pressures have tailored the same class of metabolite *via* two separate processes.

2.2.2 Insects

There are many accounts of the occurrence of prostaglandins in insects although the concentrations are in general extremely low (picogram per gram level) and their identifiication is often tentative [18]. Nevertheless, prostaglandins of the E and F groups have been reported from ten insect species representing six taxonomic orders [19]. Several insects have been studied in considerable detail and their capacity to synthesize prostaglandins is beyond doubt. The occurrence of prostaglandins in insects was first reported from the domestic cricket, *Acheta domesticus*, and homogenates of the male reproductive tract contained prostaglandin synthetase activity and could convert arachidonic acid [20:4] into prostaglandin E_2 [20]. Homogenates from various tissues of the common housefly, *Musca domestica*, have also been shown to produce PGE and PGF isomers when incubated with the appropriate eicosaenoic acids precursors [19,21]. In general the prostaglandins are implicated with the reproductive organs of insects and they appear, at least, to play a role in stimulating the release of eggs in the ovary. Intriguingly there is some evidence from *A. domesticus* to suggest that the prostaglandins, or possibly the synthetase enzyme itself, is passed from the male to the female during copulation such that the biological effect is triggered at the appropriate time [22].

2.2.3 Plants

There are about one dozen reports in the literature of prostaglandins, particularly $PGF_{2\alpha}$, occurring in plants, and in a similar manner to the insects their occurrence is almost certainly more widespread than current data indicate [23]. The purified enzyme lipoxygenase-2, one of four isoenzymes from soya beans, can convert arachidonic acid [20:4] into $PGF_{2\alpha}$, and this was the first example reported of prostaglandin biosynthesis from a non-animal source, albeit *in vitro* [24]. PGE_2 and $PGF_{2\alpha}$ (0.05–0.1%) have been isolated from the Australian red algae, *Gracilaria lichenoides*, extracts of which exhibit anti-hypertensive properties [25], and PGA_1, PGB_1, PGE_1 and PGF_1 have all been identified in onion leaves (*Allium cepa*) [26]. Interestingly $PGF_{2\alpha}$ has also been identified in a number of prokaryotic cyanobacteria the chloroplasts of which are believed to be related to plant organelles [27].

In higher plants, there is gas chromatographic–mass spectroscopy evidence to support the presence of low levels (1–2 ng/g) of $PGF_{2\alpha}$ in flowering individuals of the plant *Kalanchoe blossfeldiana* [28]. Seasonal concentrations of arachidonic acid and prostaglandins have been reported in buds of *Populus balsamifera*, and the data indicate a direct correlation between the two, implying a biosynthetic relationship [29]. Prostaglandin concentrations increase during periods of growth and decline during periods of dormancy. Most interestingly, a South African group have demonstrated that the well known inhibitors of cyclo-oxygenase, the enzyme responsible for prostaglandin biosynthesis in mammals, such as aspirin, phenylbutazone and indomethacin, inhibit the flowering of *Pharbitis nil Chois*. strain 'Violet' [30]. A connection with prostaglandin biosynthesis is not necessarily established. However the limited available evidence does raise tantalizing questions regarding the occurrence of prostaglandins in higher plants and their possible role in regulating flowering.

2.3 THE BIOSYNTHESIS OF PROSTAGLANDINS

2.3.1 Mammalian prostaglandins

In mammals Series 1, 2 and 3 of the prostaglandins are derived from eicosatrienoic [20:3], eicosatetraenoic [20:4] (arachidonic) and eicosapentaenoic [20:5] acids respectively as illustrated in Fig. 2.3. The process is the same in each case, although the level of unsaturation differs for each Series. During the transformation of the eicosaenoic acids into prostaglandins the formation of a C–C bond between C-8 and C-12 generates the characteristic cyclopentane ring system. Two (for PGA, PGB and PGC) or three (for PGD, PGE, PGF_α and PGF_β) oxygen atoms are introduced, and in Series 2 and 3 the *cis*-double bonds at C-5 and C-17 of the precursors are processed unchanged. For Series 1 the first enzyme on the pathway is fatty acid cyclo-oxygenase which mediates an oxidative cyclization to the endoperoxide intermediate PGG_1 (2.10) in Fig. 2.4. The early experiments probing this process were focused on the origin of the introduced oxygen atoms, and in his classic experiment [31] Samuelsson demonstrated that the 15(*S*)-hydroxy alcohol oxygen was derived from molecular oxygen and that both of the ring oxygen atoms of PGE_1 originate from the *same* molecule of oxygen. In the definitive experiment, 8,11,14-eicosatrienoic acid [20:3] was incubated with sheep seminal vesicles under an atmosphere of $^{16}O_2$ (46%) and $^{18}O_2$ (53%). In order to prevent the keto group of PGE_1 exchanging rapidly with water the product was reduced immediately with $NaBH_4$. Subsequent mass spectroscopy analysis of the derivatized PGE_1 indicated the presence of molecules containing either two ^{16}O or two ^{18}O isotopes attached to the ring. There was an insignificant number of molecules containing an ^{16}O ^{18}O combination, demonstrating that a single molecule of O_2 donated both ring oxygens.

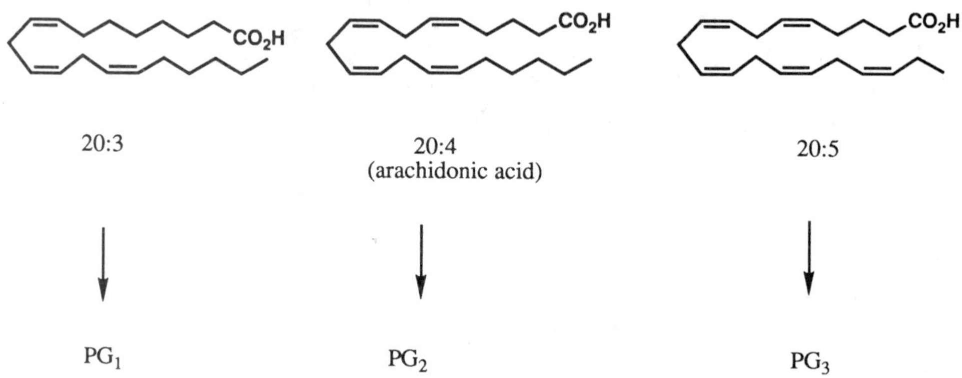

Fig. 2.3 — The eicosaenoic acid precursors in prostaglandin biosynthesis.

A series of double labelling experiments ($^{14}C/^{3}H$) was carried out by Samuelsson's group to trace the fate of individual hydrogen atoms during the conversion of 8,11,14-eicosatrienoic acid into PGE_1 [32,33]. Both the 13-*pro-R* and 13-*pro-S* hydrogen atoms were stereospecifically labelled with ^{3}H in separate experiments and the change in the $^{3}H/^{14}C$ ratio clearly demonstrated that there is a stereospecific and rate-limiting abstraction of the 13-*pro-S* hydrogen atom of the fatty acid precursor during the transformation. A mechanism was proposed as illustrated in Fig. 2.4 involving the radical abstraction of this hydrogen to generate an allylic radical at C-13 [33]. Rearrangement to a *trans* double bond with the radical centred on C-11 is followed by attack of a peroxy radical. The second oxygen atom then attacks the double bond at C-9 and initiates endoperoxide and cyclopentane ring formations, with concomitant double bond isomerization generating a *trans* allylic radical centred at C-15. Homolytic bond formation with another hydroperoxide radical generates the hydroperoxyendoperoxide species, PGG_1 (2.10). The reduced endoperoxide alcohol has been designated PGH_1 (2.11), and clearly entry into the prostaglandins requires a further reduction of the endoperoxide.

This mechanism gained considerable credibility after the isolation of PGG_2 (2.12) and PGH_2 (2.13) on incubation of arachidonic acid (Series 2 PGs) with microsomes of sheep seminal vesicles [34,35]. These compounds are unstable, with a half-life of about 5 min *in vivo*, and their characterization required a considerable level of ingenuity and skill. The enzyme cyclo-oxygenase mediates the formation of PGG_2 (2.12) and PGH_2 (2.13) [36,37]. Cyclo-oxygenase operates as a dimer with apparently identical monomeric units. It requires heme, and the cyclo-oxygenase and hydroperoxidase (reduction of the C-15 hydroperoxide) activities appear to be associated with

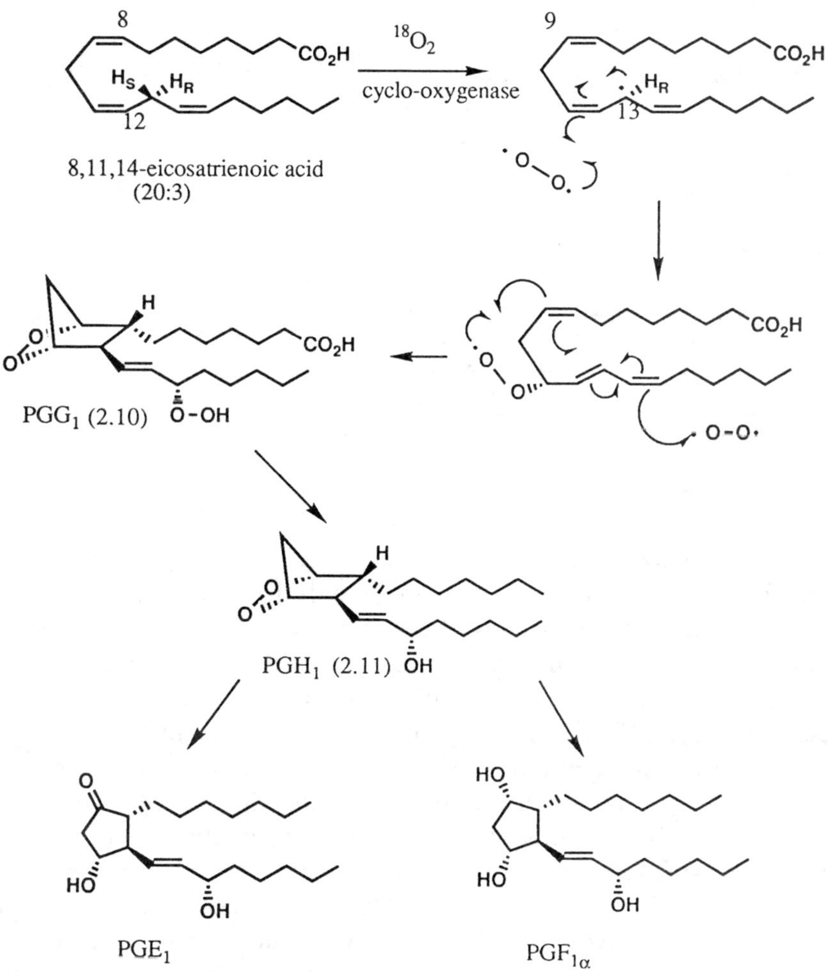

Fig. 2.4 — Mechanism of biosynthesis of the prostaglandin PG$_1$ Series.

the same protein. The D prostaglandins can arise from non-enzymatic rearrangement of PGH, although a specific D_2 isomerase activity has been identified in serum proteins and the isomerization appears to be under physiological control [38]. An endoperoxide:E isomerase converts PGH into the prostaglandins of the E group. The F group consists of a cyclopentane diol which is generated by the action of an endoperoxide:$F_{2\alpha}$ reductase. The F_2 group for example, originates from the direct reduction of the endoperoxide PGH_2, rather than from reduction of the ketone functionality of PGE_2.

arachidonic acid [20:4] —cyclo-oxygenase→

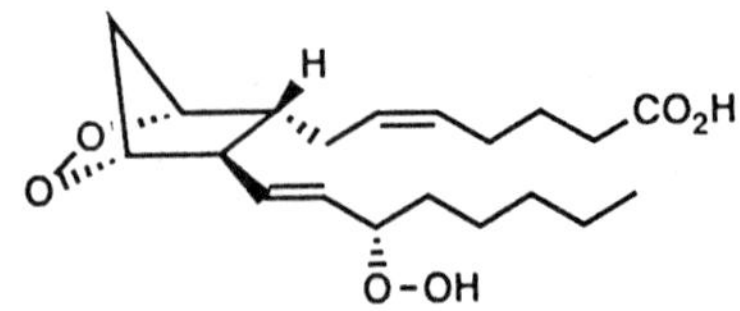

PGG_2 (2.12)

—cyclo-oxygenase→

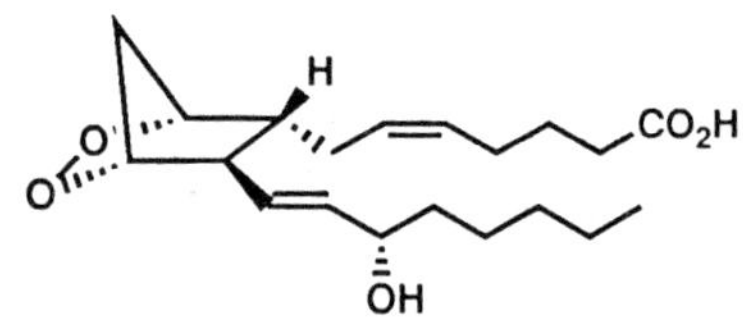

PGH_2 (2.13)

2.3.2 Marine prostaglandins

The prostaglandins found in soft corals originate from arachidonic acid but by a different pathway from that of the mammalian prostaglandins. Fruitful biosynthetic investigations have been carried out on *Clavularia viridis*, which secretes the clavulones [39]. The conversion of arachidonic acid to 8(*R*)-hydroperoxy-5(*Z*),9(*E*),11(*E*),14(*Z*)-eicosatetraenoic acid (8(*R*)HPETE) (2.14) by an acetone powder of *C. viridis* is evidence for an (8R) lipoxygenase as the initial transformation on the pathway, and the identification of pre-clavulone-A (2.17) from homogenates of *C. viridis*, incubated with arachidonic acid or 8(*R*)HPETE, clearly implicates pre-clavulone-A as a precursor to these metabolites. Incubations of 8(*R*)HPETE with acetone powders from *Plexaura homomalla* have resulted in the successful identification of allene-oxide (2.15) [40]. On the basis of these observations a biosynthetic pathway to the clavulones can be proposed which involves the conversion of 8(*R*)HPETE (2.14) to the alleneoxide (2.15) and then to an oxidopentadienyl cation (2.16). By analogy, a similar process can be predicted for the prostaglandin methyl ester acetates of *P. homomalla* such as (2.1), with alleneoxide (2.18) as an intermediate.

2.4 THROMBOXANES AND PROSTACYCLIN

Thromboxane A_2 (TXA_2) (2.19) is synthesized in the blood platelets and is very active at low concentrations at inducing platelet aggregation. Prostacyclin (PGI_2) (2.20) on the other hand is a vasodilator synthesized in blood vessel tissue (and gastric mucus) and disaggregates platelets. Both metabo-

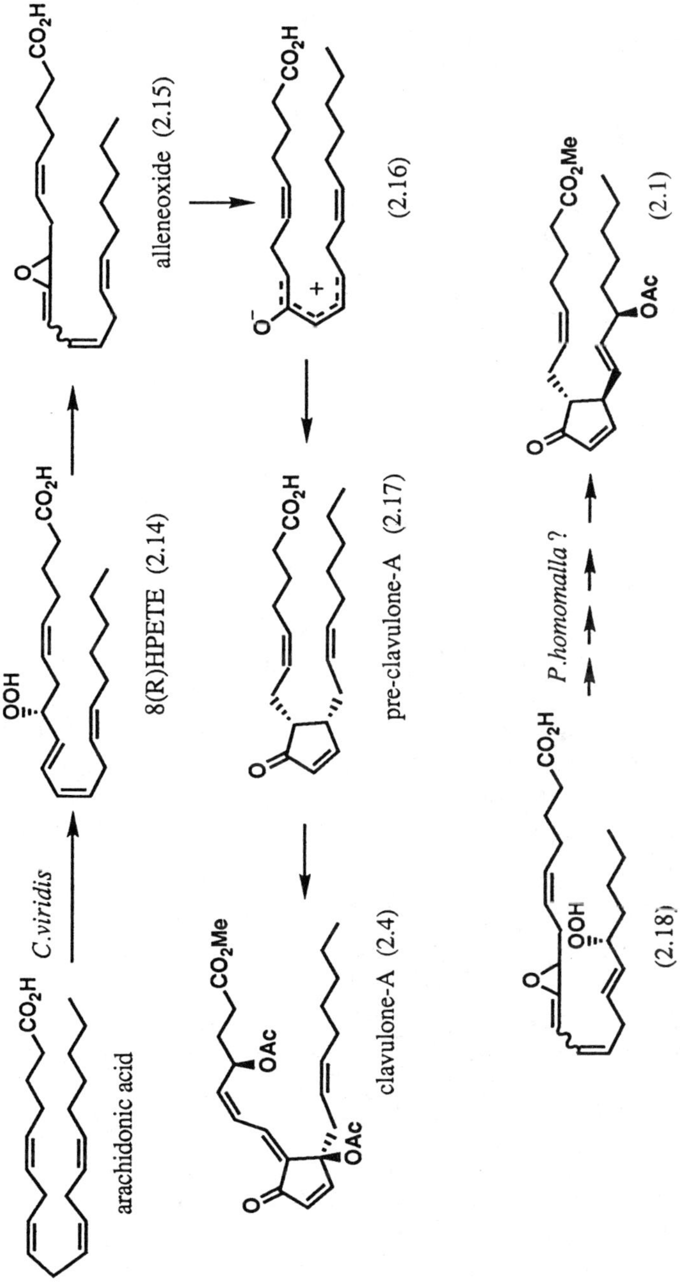
arachidonic acid
C.viridis
8(R)HPETE (2.14)
alleneoxide (2.15)
(2.16)
pre-clavulone-A (2.17)
clavulone-A (2.4)
(2.18)
P.homomalla ?
(2.1)

lites are derived from arachidonic acid and have PGH_2 (2.13) as a common biosynthetic intermediate. Thromboxane and prostacyclin synthases, the enzymes which mediate their respective transformations from PGH_2, are mechanistically very similar. The complementary activity of these hormones and their origin from the same anabolic pathway clearly indicate a sophisticated and subtle interplay controlling platelet aggregation on the surface of blood vessels.

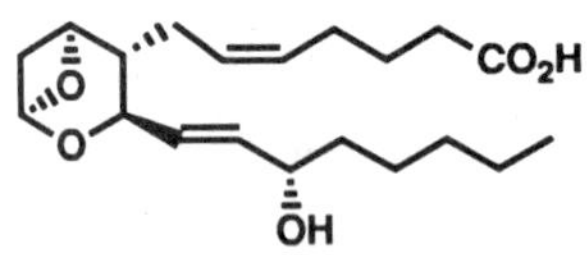

TXA_2 (2.19)

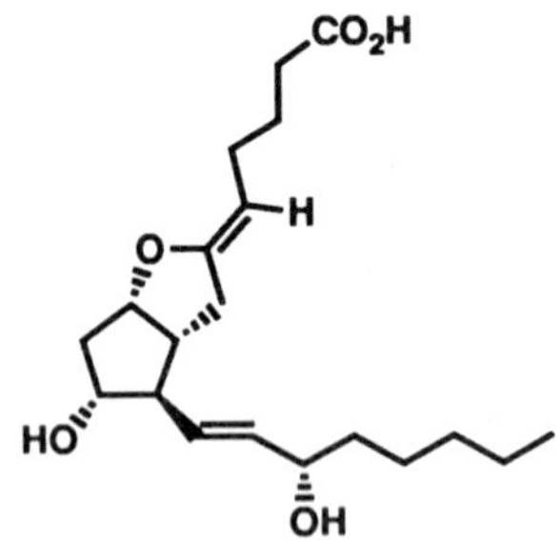

PGI_2 (2.20)

2.4.1 Thromboxanes

Of the two hormones TXA_2 (2.19) was first characterized. The initial investigations demonstrated that incubation of arachidonic acid with platelets generated an aggregating factor, the activity of which disappeared rapidly. TXA_2 is very unstable at physiological pH (half-life=30 s) as it decomposes to the biologically inactive thromboxane-B_2 (TXB_2) (2.21). The isolation of TXB_2 stimulated a series of successful trapping experiments with nucleophiles such as methanol, ethanol and azide, and in each case the corresponding thromboxane-B_2 analogue was isolated (see Fig. 2.5) [41]. This presented clear evidence for the intermediacy of an unstable electrophilic species such as TXA_2.

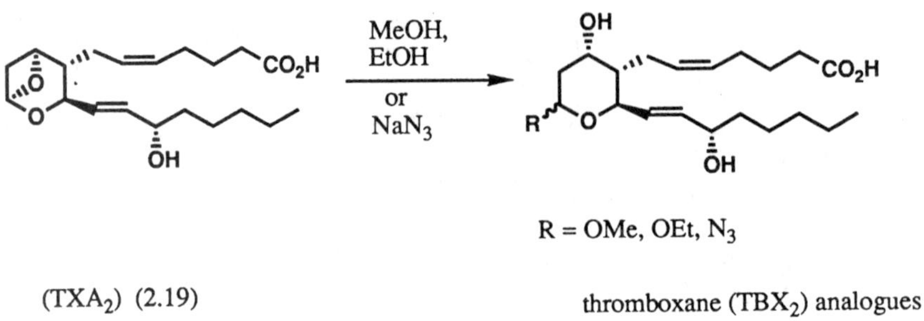

Fig. 2.5 — Trapping of TXA_2 with nucleophiles.

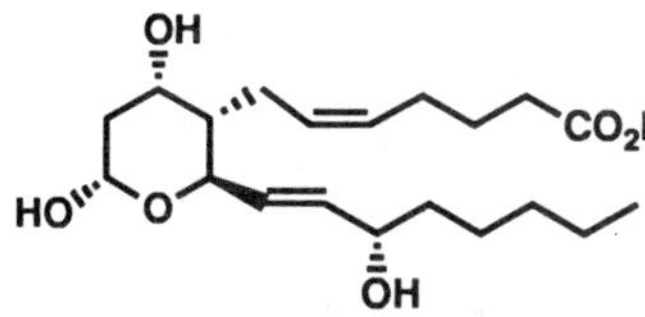

TXB$_2$ (2.21)

Two enzymes mediate the transformation of arachidonic acid to TXA$_2$. Fatty acid cyclo-oxygenase generates PGH$_2$, and then thromboxane-A$_2$ synthase controls the isomerization to TXA$_2$. Thromboxane synthase has been purified from several sources and shown to be a cytochrome P$_{450}$ enzyme containing an active site heme co-ordinated to a thiolate ligand. Fig. 2.6 illustrates a mechanism for thromboxane synthase consistent with our

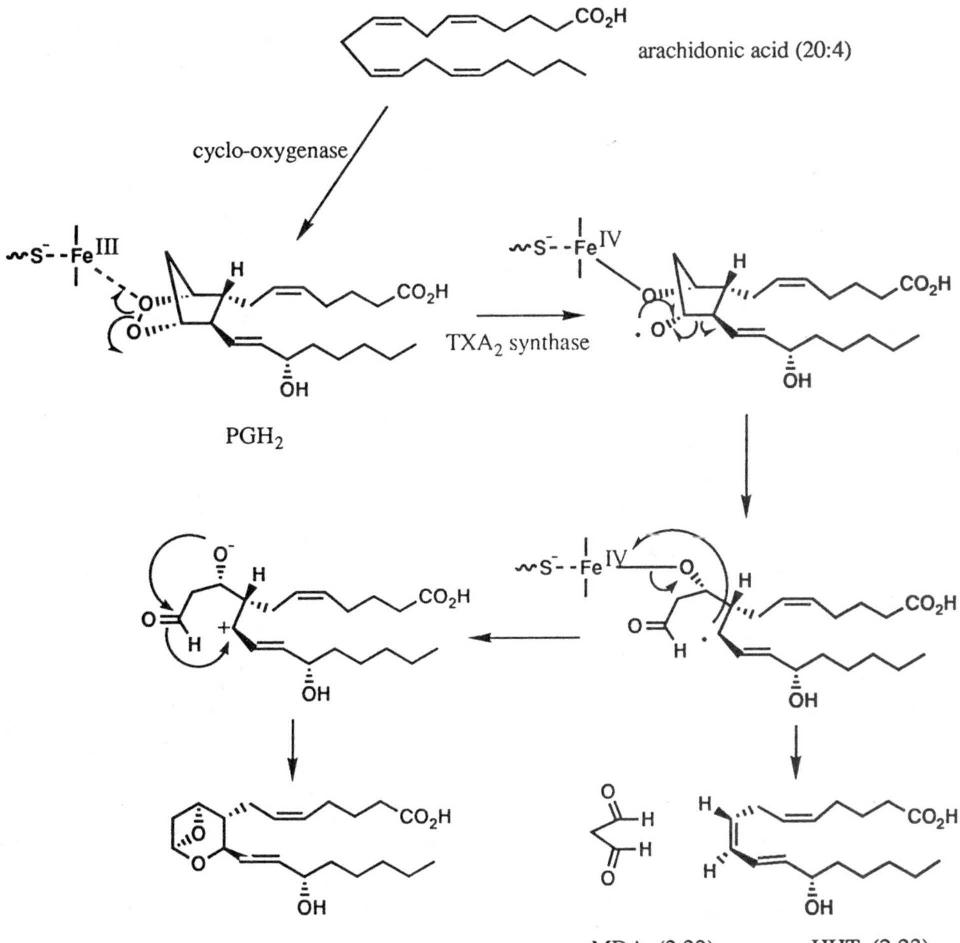

Fig. 2.6.

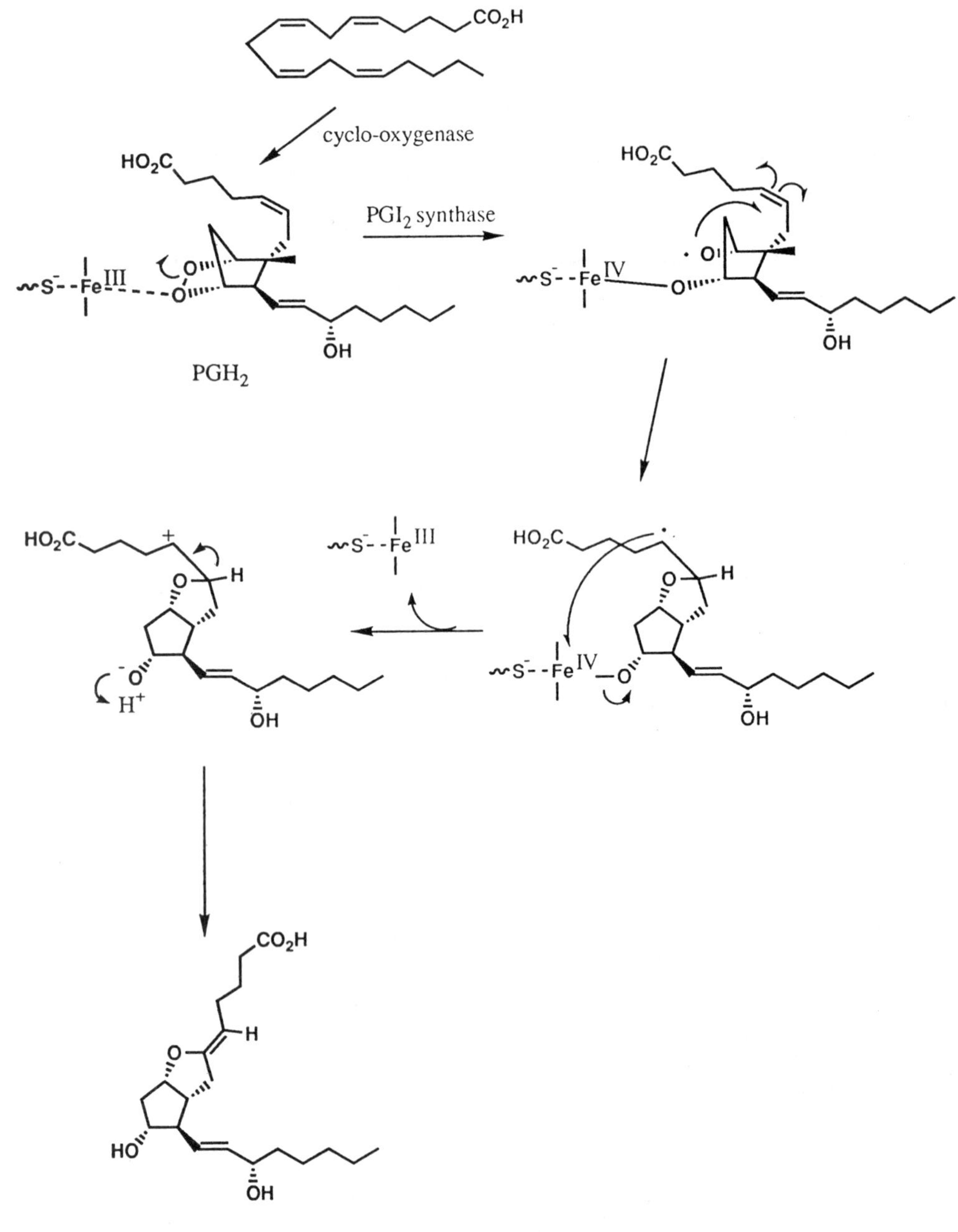

PGI$_2$ (2.20)

Fig. 2.7 — A mechanistic rationale for prostacyclin biosynthesis from PGH$_2$.

current understanding of this enzyme system [42]. Interestingly, thromboxane synthase mediates the conversion of PGH_2 into malonic dialdehyde (MDA) (2.22) and 12*(S)*-hydroxyheptadeca-5*Z*,8*E*,10*E*-trienoic acid (HHT) (2.23) in a competing reaction to TXA_2 synthesis [43]. The rates of these reactions are similar with equimolar amounts of TXA_2 and HHT produced, and they almost certainly originate from a common intermediate in the process. The role of MDA and HHT is unclear but it is unlikely that they are just passive byproducts of this reaction; most probably they exert some local physiological response.

2.4.2 Prostacyclin

Prostacyclin (PGI_2) (2.20) is the primary product of arachidonic acid metabolism in vascular tissue, and its presence was pursued as a consequence of a platelet anti-aggregating activity in various bioassays when assessing the decomposition of PGH_2 [44]. PGI_2 decomposes (half-life 2 min) under physiological conditions to give 6-ketoprostaglandin-$F_{1\alpha}$, but is stable in acetone at low temperatures. The structure of PGI_2 was elucidated largely by its synthesis from $PGE_{2\alpha}$ methyl ester, isolated from the soft coral *P. homomalla*, and then by comparison of the biological activity of the synthetic material with the natural [45]. PGI_2 synthase has been purified and posesses very similar characteristics to TXA_2 synthase. In the PGI_2 system the C-11 peroxide oxygen co-ordinates to iron as opposed to the C-9 oxygen in TXA_2 synthase, and it would appear that it is only this subtle control on regioselectivity that distinguishes the two isomerization processes [42] (see Fig. 2.7).

2.4.3 Levuglandins

It can be seen from the above discussion that the endoperoxide PGH_2 (2.13) produced by cyclo-oxygenase activity on arachidonic acid plays a central role in the production of the prostanoid hormones. The multitude of breakdown pathways of the endoperoxide moiety allows entry into the prostaglandins, thromboxanes and prostacyclin, and it is interesting at this point to speculate on another class of prostanoid hormones which may exist *in vivo*, derived from PGH_2. A series of studies monitoring the breakdown of PGH_2 in dimethylsulphoxide (DMSO) at 37°C have revealed that under these conditions PGH_2 gives rise to two isomeric aldehydes (2.24) and (2.25) in 70–80% yield, and in buffered aqueous solution in 22% yield [46]. These compounds have been designated the levuglandins LGE_2 and LGD_2, and they arise from yet another rearrangement of the endoperoxide moiety of PGH_2. Their ease of formation *in vitro* is exciting and suggests that they may ultimately emerge as true *in vivo* metabolites. *In vitro* studies with synthetic

LGE_2 have demonstrated that this compound cross-links proteins, although the biological consequences of this are unclear. LGE_2 has also been shown to inhibit cell division of fertilized sea urchin eggs [47].

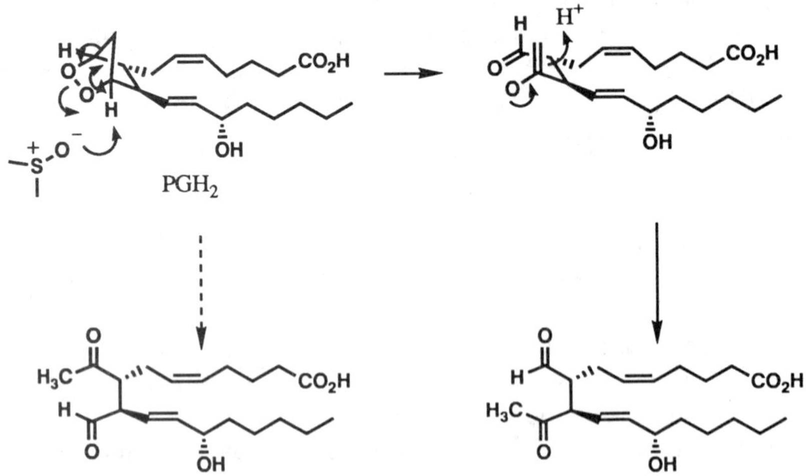

LGE_2 (2.24) LGD_2 (2.25)

2.5 LEUKOTRIENES

A series of metabolites of polyunsaturated fatty acids (eicosaenoic acids) are biosynthesized in leukocytes (white blood cells) and are responsible for various aspects of the inflammation response. The major metabolites are generated from a pathway initiated by a lipoxygenase-catalysed reaction on arachidonic acid, and they have been termed the *leukotrienes* owing to their occurrence in these cells. The first identified compound of this group was the 5(*S*)-hydroxyeicosatetraenoic acid (5-HETE) (2.27) which arises from the reduction of the corresponding hydroperoxide 5-HPETE (2.26), the product of the lipoxygenase reaction on arachidonic acid. The 5(*S*), 12(*R*)-dihydroxy derivative, leukotriene-B_4 (LTB_4) (2.29), displays a variety of physiological activities and is arguably the most significant of the series. It too is derived from 5-HPETE, although not from 5-HETE, and appears to arise via 5,6-epoxyeicosatetraenoic acid (LTA_4) (2.28). Stereochemical studies have revealed that the 10-*pro-R* hydrogen of 5-HPETE is removed during leukotriene biosynthesis. Generation of the epoxide functionality can be rationalized by the dehydration illustrated in Fig. 2.8, a process compatible with experimental observation. The C-5 oxygen atom of LTB_4 is derived from molecular oxygen and the oxygen at C-12 from the aqueous medium. In a similar strategy to the thromboxane studies it became pertinent to ask whether appropriate nucleophiles could intercept the biosynthetic process and generate LTB_4 analogues. Significantly, when incubations were quenched with methanol the corresponding derivative (2.30) was isolated. This clearly points to an electrophilic intermediate, such as LTA_4, being trapped by a nucleophile at the terminus of the conjugated system, with concomitant double bond rearrangement and ring opening.

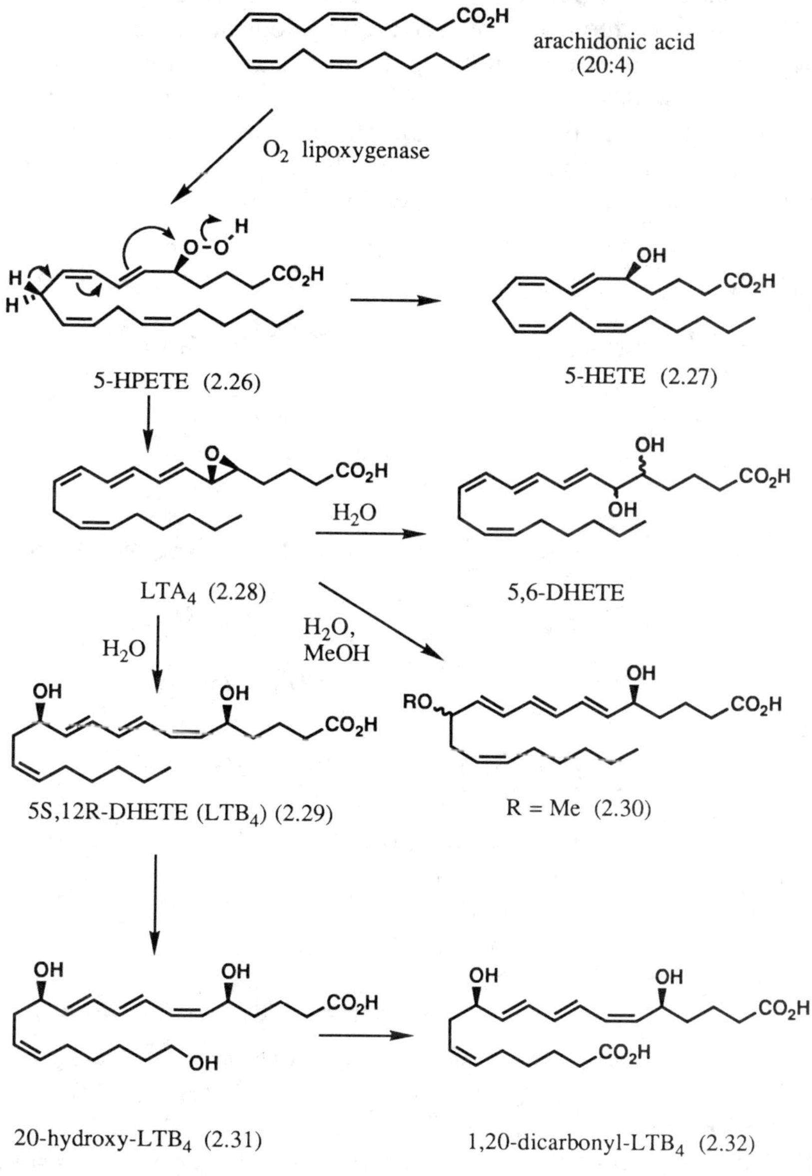

Fig. 2.8.

Two further metabolites of LTB_4 have been identified in human leukocytes after incubation with arachidonic acid and stimulation with a chemotactic peptide. In this study C-20 hydroxyl-LTB_4 (2.31) was the minor compound and the major metabolite was the fully oxidized 1,20-dicarboxyl-LTB_4 (2.32) [48].

Leukotriene-A_4 (LTA_4) (2.28) is an intermediate in a series of metabolites known as the slow reacting substances of anaphylaxis (SRS-A). These compounds are secreted in response to immunological challenge on mammalian leukocytes. They are partially lipid and partially peptide in nature and are typified by leukotrienes-C_4 to -F_4 (2.33–2.36) which originate after nucleophilic attack of the thiol group of glutathione at C-6 of the LTA_4 epoxide. A series of these compounds have been identified and their biosynthetic relationships are illustrated in Fig. 2.9.

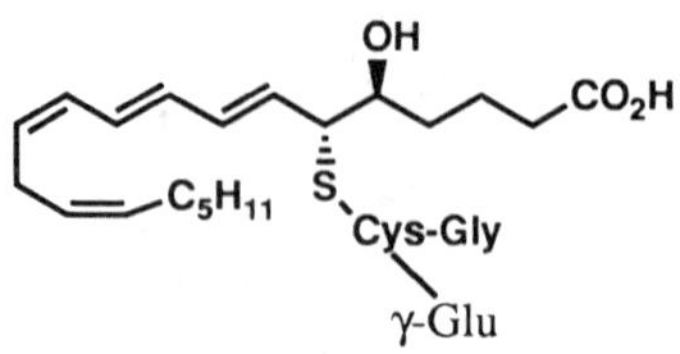

LTC_4 (2.33)

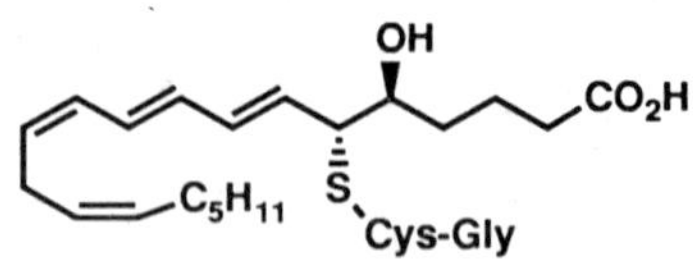

LTD_4 (2.34)

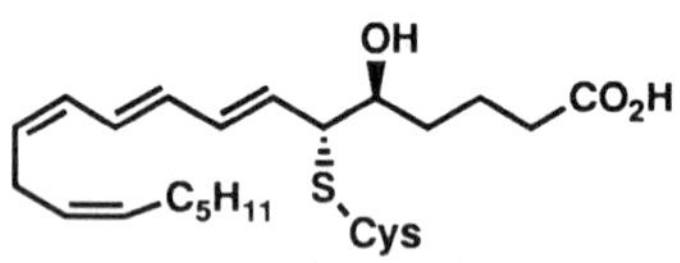

LTE_4 (2.35)

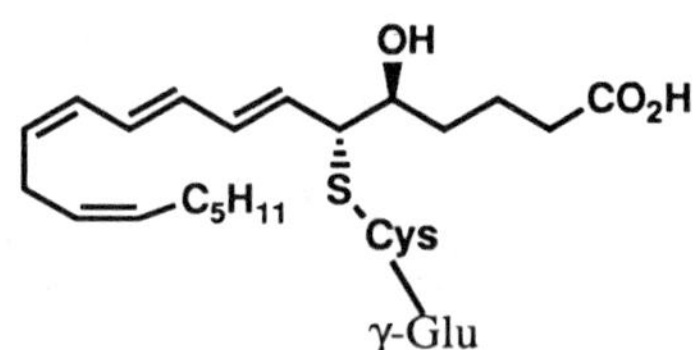

LTF_4 (2.36)

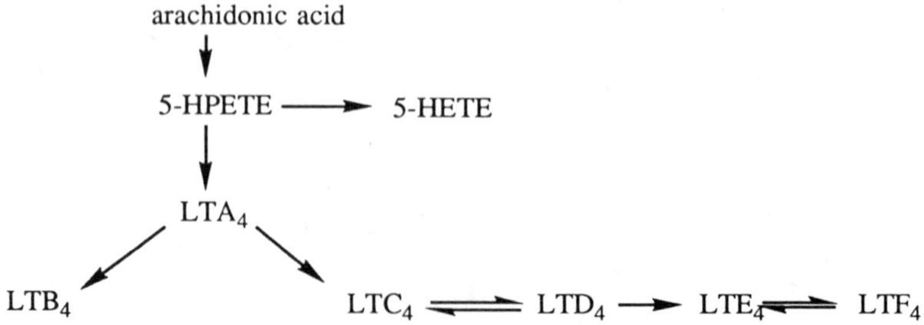

Fig. 2.9 — The structures and the biosynthetic relationships of the SRS-A.

REFERENCES

[1] R. Kurzrok and C. C. Leib, Biochemical studies on human semen II. The action of semen on the human uterus. *Proc. Soc. Exp. Biol. Med.*, 1930, **28**, 268.

[2] M. W. Goldblatt, A depressor substance in seminal fluid. *J. Soc. Chem. Ind. London*, 1933, **52**, 1056.
[3] M. W. Goldblatt, Properties of human seminal plasma. *J. Physiol.* (London), 1935, **81**, 208.
[4] U. S. von Euler, A depressor substance in the vesicular gland. *J. Physiol.* (London), 1935, **84**, 21P.
[5] U. S. von Euler, On the specific vasodilating and pain muscle stimulating substances from accessory genital glands in man and certain animals (prostaglandin and vesiglandin). *J. Physiol.* (London), 1936, **88**, 213.
[6] U. S. von Euler and R. Eliasson, *Prostaglandins*, Academic Press, 1967, p. 50.
[7] S. Bergstrom, The prostaglandins: from the laboratory to the clinic (Nobel lecture). *Angew. Chem. Int. Ed. Engl.*, 1983, **22**, 858.
[8] D. H. Nugteren, D. A. van Dorp, S. Bergstrom, M. Hamberg and B. Samuelsson, Absolute configuration of the prostaglandins. *Nature*, (London), 1966, **212**, 38.
[9] J. B. Smith, M. J. Silver, C. M. Ingerman and J. J. Kocsis, Prostaglandin D_2 inhibits the aggregation of human platelets. *Thromb. Res.*, 1974, **5**, 291.
[10] M. A. Wasserman, D. W. DuCharm, R. L. Griffin, G. L. Degraff and F. G. Robinson, Bronchopulmonary and cardiovascular effects of prostaglandin D_2 in the dog. *Prostaglandins*, 1977, **13**, 255.
[11] A. J. Weinheimer and R. L. Spraggins, The occurrence of two new prostaglandin derivatives in the gorgonian *Plexaura homomalla*. Chemistry of coelenterates XV. *Tetrahedron Lett.*, 1969, **5**, 185.
[12] W. P. Schneider, R. D. Hamilton and L. E. Rhuland, Occurrence of esters of (15*S*)-prostaglandin-A_2 and E_2 in coral. *J. Am. Chem. Soc.*, 1972, **94**, 2122, 2124.
[13] M. Kobayashi, T. Yasuzawa, M. Yoshihara, H. Akutsu, Y. Kyogoku and I. Kitagawa, Four new prostanoids: claviridenone-A, -B, -C and -D from Okinawan soft coral *Clavularia viridis*. *Tetrahedron Lett.*, 1982, 5331.
[14] I. Kitagawa, M. Kobayashi, T. Yasuzawa, B. W. Son, M. Yoshihara and Y. Kyogoku, New prostanoids from soft coral. *Tetrahedron*, 1985, **41**, 995.
[15] H. Kikuchi, Y. Tsukitani, K. Iguchi and Y. Yamada, Clavulones: new type of prostanoids from the stolonifer *Clavularia viridis* quoy and gaimard. *Tetrahedron Lett.*, 1982, 5171.
[16] Y. Komoda, T. Kanayasu and M. Ishikawa, Prostaglandin $F_{2\alpha}$ from the Japanese coastal gorgonian, *Euplexaura erecta*. *Chem. Pharm. Bull.*, 1979, **27**, 2491.

[17] S. Cormley, Y. Kashman, Y. Loya and Y. Benayahu, New prostaglandin (PGF) derivatives from the soft coral *Lobophyton depressum. Tetrahedron Lett.*, 1980, **21**, 875.
[18] U. E. Brady, Review: prostaglandins in insects. *Insect Biochem.*, 1983, **13**, 443.
[19] E. J. Wakayama, J. W. Dillwith and G. J. Blomquist, Characterization of prostaglandin synthesis in the housefly, *Musca domestica* (L). *Insect Biochem.*, 1986, **16**, 903.
[20] D. B. Destephano, U. E. Brady and R. E. Lovins, Synthesis of prostaglandin by reproductive tissue of the male house cricket, *Acheta domesticus, Prostaglandins*, 1974, **6**, 71.
[21] E. J. Wakayama, J. W. Dillwith and G. J. Blomquist, Occurrence and metabolism of prostaglandins in the housefly, *Musca domestica* (L.). *Insect Biochem.*, 1986, **16**, 895.
[22] D. B. Destephano, U., E. Brady and C. A. Farr, Factors influencing oviposition behaviour in the cricket, *Acheta domesticus. Ann. Ent. Soc. Am.*, 1982, **75**, 111.
[23] E. G. Groenewald, G. H. J. Kruger, H. de. Wet, P. J. Botes and A.-J. van der Westhuizen, Detection of prostaglandin $F_{2\alpha}$ by radioimmunoassay in three South African bryophytes and its relation to the occurrence of polyunsaturated fatty acids. *South African J. Sci.,* 1990, **86,** 152.
[24] G. S. Bild, S. G. Bhat, C. S. Ramadoss and B. Axelrod, Biosynthesis of a prostaglandin by a plant enzyme. *J. Biol. Chem.,* 1978, **253**, 21.
[25] R. P. Gregson, J. F. Marwood and R. J. Quinn, The occurrence of prostaglandins PGE_2 and $PGF_{2\alpha}$ in a plant. The red alga *Gracilaria lichenoids. Tetrahedron Lett.*, 1979, 4505.
[26] S. Al-Nagdy, M. O. Abdel Rahman and H. I. Heiba, Extraction and identification of different prostaglandins in *Allium cepa. Comp. Biochem. Physiol.*, C, 1986, **85**, 163.
[27] G. H. J. Kruger, E. G. Groenewald, H. de Wet and P. J. Botes, The occurrence of prostaglandin $F_{2\alpha}$ in procaryotic organisms. *S.-Afr. Tydskr. Plantk.*, 1990, **56**, 150.
[28] B. Janistyn, Gas chromatographic–mass spectroscopic identification of prostaglandin $F_{2\alpha}$ in the flowering plant *Kalanchoe blossfediana* v. Pollen. *Planta*, 1982, **154**, 485.
[29] E. D. Levin, E. V. Isaeva and V. E. Cherepanova, Arachidonic acid and prostaglandins in buds of *Populus balsamifera. Phytochemistry*, 1990, **29**, 2325.
[30] E. G. Groenewald and J. H. Visser, The effect of certain inhibitors of prostaglandin biosynthesis on flowering *Pharbitis nil. Z. Pflanezenphysiol. Bd.*, 1974, **71S**, 67.
[31] B. Samuelsson, On the incorporation of oxygen in the conversion of

8,11,14-eicosatrienoic acid to prostaglandin E. *J. Am. Chem. Soc.*, 1965, **87**, 3011.

[32] D. Klenberg and B. Samuelsson, The biosynthesis of prostaglandins E_1 studied with specifically ^{3}H-labelled 8,11,14-eicosatrienoic acids. *Acta Chem. Scand.*, 1965, **19**, 534.

[33] M. Hamberg and B. Samuelsson, On the mechanism of the biosynthesis of prostaglandins E_1 and $E_{1\alpha}$. *J. Biol. Chem.*, 1967, **242**, 5336.

[34] M. Hamberg and B. Samuelsson, detection and isolation of the endoperoxide intermediate in prostaglandin biosynthesis. *Proc. Nat. Acad. Sci. USA.*, 1973, **70**, 899.

[35] D. H. Nugteren and E. Hazelhof, Isolation and properties of intermediates in prostaglandin biosynthesis. *Biochem. Biophys. Acta*, 1973, **326**, 448.

[36] M. Hemler, W. E. M. Lands and W. Smith, Purification of the cyclo-oxygenase that forms prostaglandins. Demonstration of the two forms of iron in the holoenzyme. *J. Biol. Chem.*, 1976, **251**, 5575.

[37] T. Miyamoto, S. Yamamoto and O. Hayaishi, Prostaglandin synthetase system. Resolution into oxygenase and isomerase components. *Proc. Nat. Acad. Sci.*, 1974, **71**, 3645.

[38] T. Wattanabe, S. Narumiya, T. Shimizu and O. Hayaishi, Characterisation of the biosynthetic pathway of prostaglandin D_2 in human platelet rich plasma. *J. Biol. Chem.*, 1982, **257**, 14847.

[39] E. J. Corey, M. d'Alarcao, S. P. T. Matsuda, P. T. Lansbury Jr. and Y. Yamada, Intermediacy of 8-(*R*)-HPETE in the conversion of arachidonic acid to pre-clavulone A by *Clavularia viridis*. Implications for the biosynthesis of marine prostanoids. *J. Am. Chem. Soc.*, 1987, **109**, 289.

[40] A. R. Brash, Formation of an alleneoxide from (8*R*)-8-hydroperoxyeicosatetraenoic acid in the coral *Plexaura homomalla*, *J. Am. Chem. Soc.*, 1989, **111**, 1891.

[41] M. Hamberg, J. Svensson and B. Samuelsson, Thromboxanes: a new group of biologically active compounds derived from prostaglandin endoperoxides. *Proc. Nat. Acad. Sci. USA.*, 1975, **72**, 2994.

[42] M. Hecker and V. Ullrich, On the mechanism of prostacyclin and thromboxane A_2 biosynthesis. *J. Biol. Chem.*, 1989, **264**, 141.

[43] U. Diczfalusy, P. Falardeau and S. Hammarstrom, Conversion of prostaglandin endoperoxides to C_{17}-hydroxyacids catalysed by human platelet thromboxane synthetase. *FEBS Letts.*, 1977, **84**, 271.

[44] R. J. Vane, Adventures and excursions in bioassay: the stepping stone to prostacyclin (Nobel lecture). *Angew. Chem. Int. Ed. Engl.*, 1983, **22**, 741.

[45] R. A. Johnson, F. H. Lincon, J. L. Thompson, E. G. Nidy, S. A. Mizsak and U. Axen, Synthesis and stereochemistry of prostacyclin and

the synthesis of 6-ketoprostaglandin $F_{1\alpha}$. *J. Am. Chem. Soc.*, 1977, **99**, 4182.

[46] R. G. Salomon, D. B. Miller, M. G. Zagorski and D. J. Coughlin, Solvent induced fragmentation of prostaglandin endoperoxides. New aldehyde products from PGH_2 and a novel intramolecular 1,2-hydride shift during endoperoxide fragmentation in aqueous solution. *J. Am. Chem. Soc.*, 1984, **106**, 6049.

[47] K. K. Marthi, R. G. Salomon and H. Sternlicht, Levuglandin E_2 inhibits mitosis and microtubule assembly. *Prostaglandins*, 1990, **39**, 611.

[48] W. Jubitz, O. Radmark, C. Malmsten, G. Hansson, J. A. Lindgren, J. Palmblad, A.-M. Uden and B. Samuelsson, A novel leukotriene produced by stimulation of leukocytes with formylmethionylleucylphenylanine. *J. Biol. Chem.*, 1982, **257**, 6106.

3

Fungal and plant polyketides

The vast majority of the polyketide secondary metabolites identified so far in Nature are products of the fungi. Representatives from all phyla elaborate polyketides but they are ubiquitous in the fungi imperfecti (or Deuteromycota) — lower fungi which lack organs for sexual reproduction — and the majority of polyketides are derived from this source.

Plants are less prolific producers of polyketides and tend to generate simple aromatic phenols by this route, although a number of plant alkaloids are emerging which are true acetogenins (see section 3.5.3). The more elaborate plant aromatics are, in general, derived from metabolites of the shikimate pathway (see section 3.2.3). Many of these, such as the important flavonoid and related metabolites, arise from a combination of shikimate and polyketide pathways (e.g. sections 3.2.5 and 3.5.5), but relatively few are entirely polyketide in origin.

The majority of polyketide secondary metabolites identified from fungal and plant sources are aromatic, and although non-aromatic compounds exist they are more rare. The predominance of aromatic compounds may be influenced by their ease of identification and isolation; however, the large number of such compounds isolated, forces the conclusion that most of the polyketide synthases are optimized for the production of aromatic compounds.

Birch and Donovan suggested in their seminal paper outlining the 'acetate hypothesis' [1] that the folding of poly-β-keto-thioesters, generated by the head to tail assembly of acetyl units, could account for the common phenols and many other aromatic species which are distributed widely in Nature. Implicit in this analysis is the absence of a β-ketoacyl reductase activity acting ritually on the β-ketone functionality after each condensation event, and clearly the complete absence, or only partial deployment, of the reductase activity by polyketide synthases emerges as an important difference from fatty acid biosynthesis. One of the first biosynthetic investigations used to substantiate the 'acetate hypothesis' was carried out on the antibiotic griseofulvin (3.1) from the fungi *Penicillium griseofulvin* [2]. Birch demonstrated the efficient incorporation of [1-^{14}C]-acetate into the carbon skeleton of griseofulvin (3.1) and was then able to establish a labelling pattern consistent with the folding of a heptaketide precursor. Since then, numerous biosynthetic experiments have borne out the validity of the 'acetate hypothesis', and the application of the analysis has provided a useful biogenetic tool for firming up the structures of many polyketide metabolites.

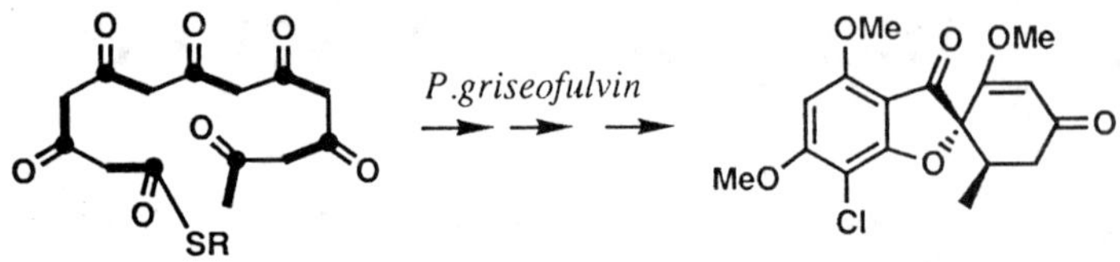

(3.1)

3.1 SINGLE AROMATIC RINGS

The process of polyketide chain assembly has analogies with fatty acid biosynthesis. Acetyl-CoA is employed as a starter unit and a series of decarboxylative condensations, utilizing malonyl-CoA, generates an extended poly-β-keto-thioester as shown opposite. Unlike fatty acid biosynthesis, however, the β-ketone functionality is not always reduced, and the various polyketide synthases can hold the assembled poly-β-keto intermediate and then mediate cyclizations to specific aromatic ring systems. The poly-β-keto-thioesters are inherently unstable, as they possess active methylene groups proximate to ketones or thioesters and are therefore susceptible to a variety of intramolecular condensations. It is rational to suppose that the poly-β-keto functionality is stabilized on the synthase, possibly by metal ion chelation, and that specific cyclization modes are dictated by the spatial constraints of individual synthases. In two of the simplest cases the common fungal metabolites orsellinic acid (3.3) and acetylphloroglucinol (3.4) are derived from the condensation of an acetyl-CoA starter unit followed by three malonyl-CoA condensations (tetraketides). The resultant poly-β-keto-thioester (3.2) cyclizes via an aldol-type condensation and then undergoes a dehydration and enolizations to give orsellinic acid (3.3) after hydrolysis. An intramolecular C-acylation of (3.2), rather than an aldol condensation, generates (3.4) after appropriate enolizations and hydrolysis. In both of these cases there is no β-keto reductase involved. This is by no means universal and many polyketide synthases do possess such activity. For example, the absence of a C-4 oxygen in 6-methylsalicylic acid (3.5) is due to carbonyl reduction by a NADPH-dependant β-keto reductase, followed by dehydration to a *cis* double bond. The origin of orcinol (3.6) and *m*-cresol (3.7) are rationalized by decarboxylation of orsellinic and 6-methylsalicylic acids respectively.

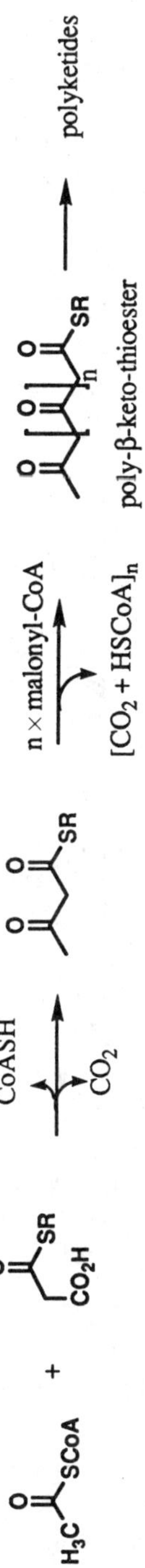
H3C SCoA
SR
CO2H
CoASH
CO2
SR
n × malonyl-CoA
[CO2 + HSCoA]n
SR
poly-β-keto-thioester
polyketides

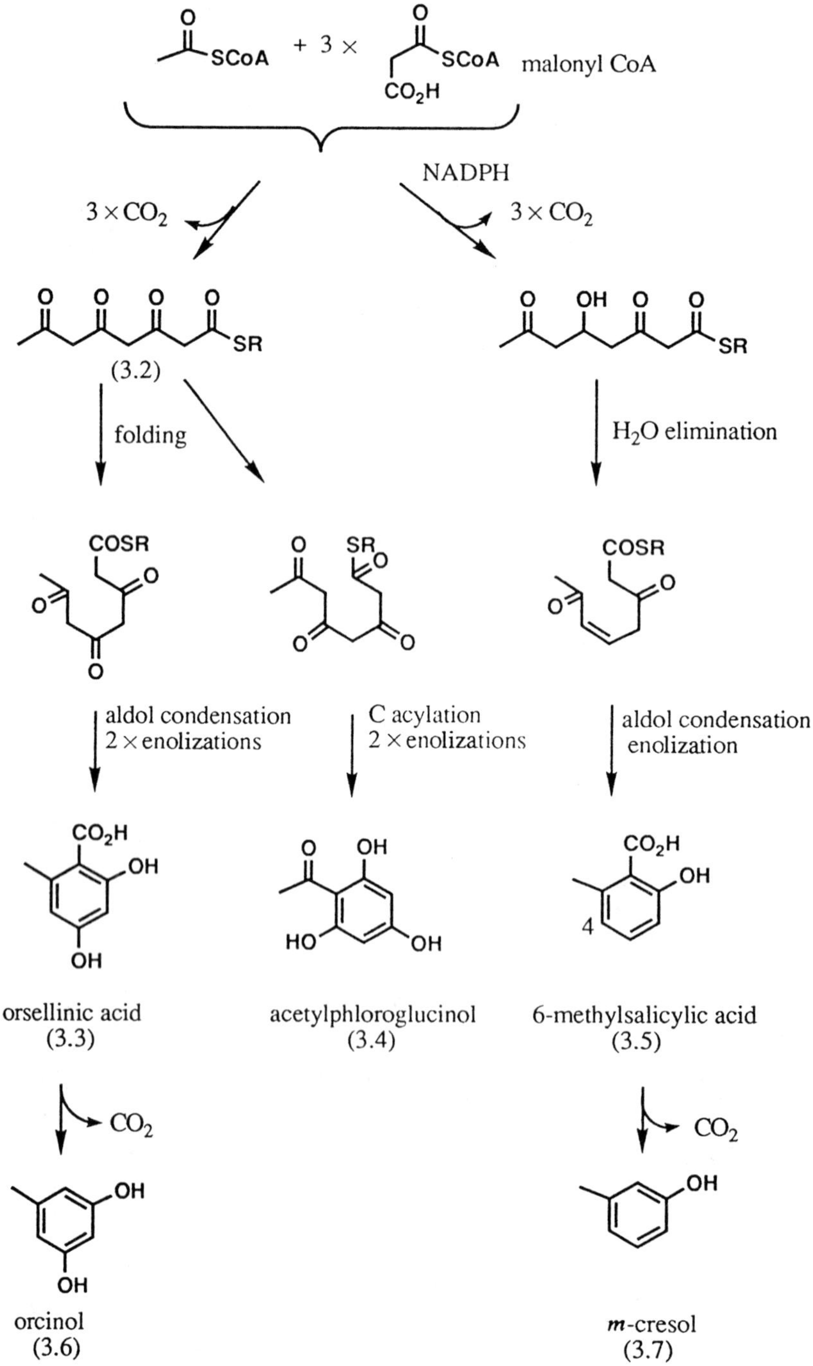
SCoA
+ 3 ×
SCoA
CO2H
malonyl CoA
NADPH
3 × CO2
3 × CO2
SR
(3.2)
SR
folding
H2O elimination
COSR
SR
COSR
aldol condensation
2 × enolizations
C acylation
2 × enolizations
aldol condensation
enolization
CO2H
OH
OH
OH
HO
OH
CO2H
OH
4
orsellinic acid
(3.3)
acetylphloroglucinol
(3.4)
6-methylsalicylic acid
(3.5)
CO2
CO2
OH
OH
OH
orcinol
(3.6)
m-cresol
(3.7)

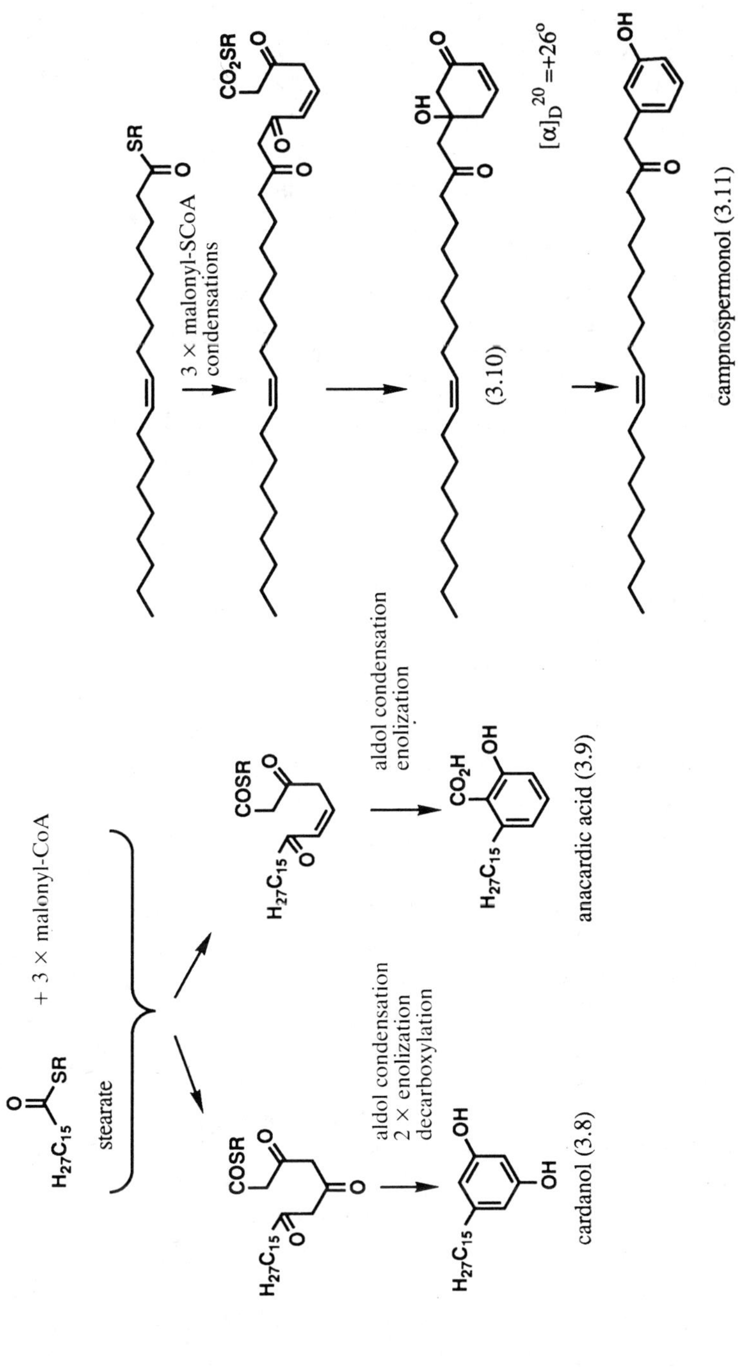
stearate
+ 3 × malonyl-CoA
aldol condensation
2 × enolization
decarboxylation
cardanol (3.8)
aldol condensation
enolization
anacardic acid (3.9)
3 × malonyl-SCoA
condensations
(3.10)
$[\alpha]_D^{20}$ =+26°
campnospermonol (3.11)

Cardanol (3.8) and anacardic acid (3.9), isolated from the oil of cashew nuts, have long aliphatic chains attached to aromatic nuceli and are clearly derived from a C_{16}-stearate starter unit by directly analogous processes to orcinol (3.6) and 6-methylsalicylic acid (3.5). Campnospermonol (3.11) from tigaso oil, by analogy, utilizes oleate as a primer for four malonate condensations and is processed in a similar manner to *m*-cresol (3.7). The optically active compound (3.10) is co-produced with campnospermonol and for this reason has been implicated as a biosynthetic precursor [3].

Those polyketides which possess single aromatic ring systems will logically have similar modes of assembly. The biosyntheses of 6-methylsalicylic acid and orsellinic acid have received considerable attention, and it is appropriate at this stage to summarize some of these results, as they have specific relevance to aromatic polyketides in general. There is, however, an apparent difference in stereochemical control between the assembly processes in these cases which points to a possible variation in the degree of enzymatic control in going from one system to another.

3.1.1 6-Methylsalicylic acid

In 1970 an enzyme extract from *Penicillium patulum* was isolated and extensively purified [5]. The extract converted acetyl-CoA and malonyl-CoA into 6-methylsalicylic acid (6-MSA) in the presence of NADPH. 6-MSA synthase is a single multifunctional protein characteristic of the more highly evolved Type I mammalian fatty acid synthases (see section 4.8.3). Both systems utilize the same substrates and transesterify them onto 4′-phosphopantetheinyl thiol residues. Each is inhibited by iodoacetamide and *N*-ethylmalinimide which is consistent with free thiol groups being important for activity. A scheme detailing the individual steps in 6-MSA biosynthesis is outlined in Fig. 3.1. There are 11 transformations catalysed by seven catalytic regions (domains) of the multifunctional protein as well as two intramolecular transacylations. The seven activities are an acetyl transferase, a malonyl transferase, a condensing enzyme, a β-keto reductase, a dehydratase, a cyclase and a deacylase. By comparison with FAS, 6-MSA synthase has an additional cyclization activity where an aldol condensation generates the ring system, but it is devoid of an enoyl reductase activity.

An interesting feature of 6-MSA synthase is the generation of triacetic acid lactone (TAL) when acetyl-CoA and malonyl-CoA are incubated in the *absence* of NADPH. This has been interpreted as evidence supporting the reduction of a C_6 intermediate prior to the final malonyl condensation [5]. The production of TAL in the absence of NADPH is, however, not unique to

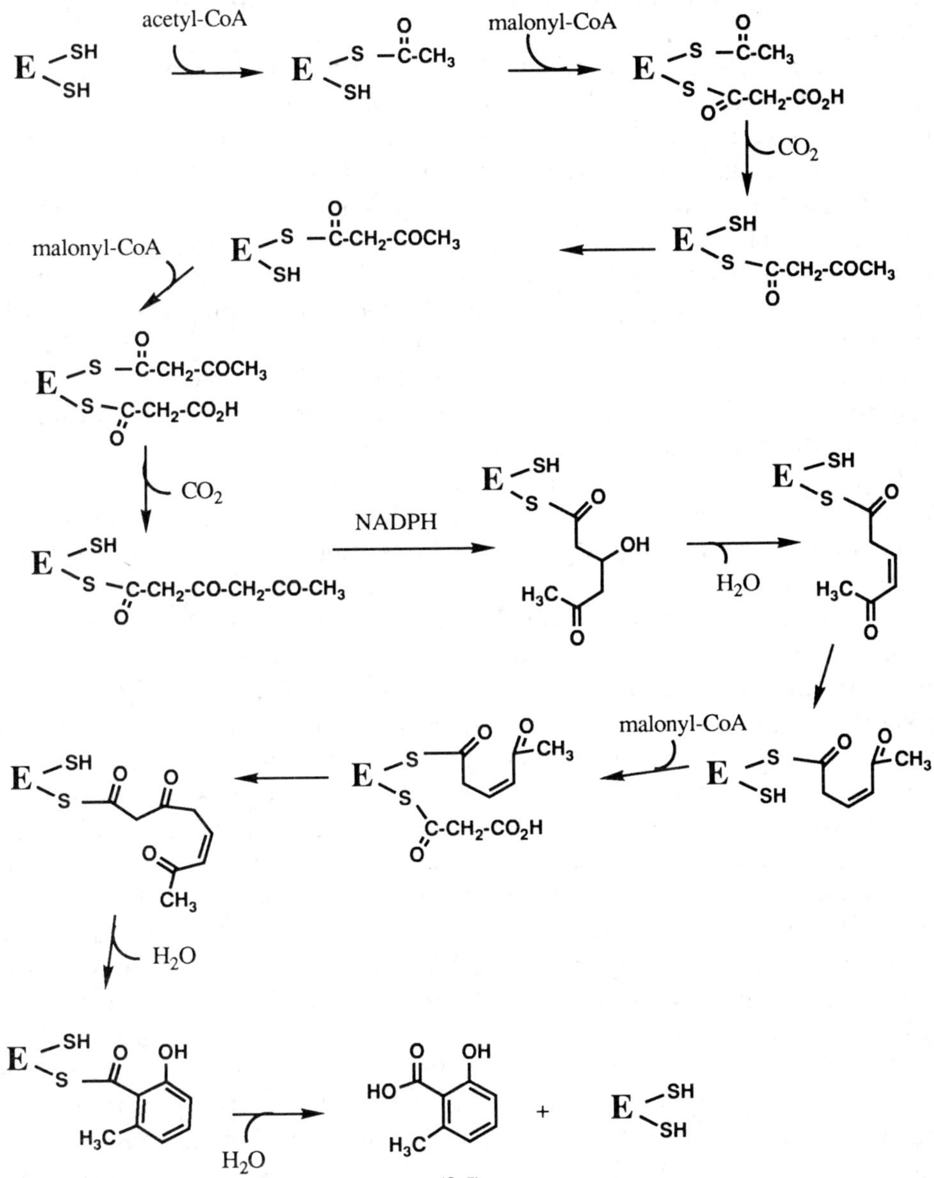
acetyl-CoA
malonyl-CoA
CO_2
malonyl-CoA
CO_2
NADPH
H_2O
malonyl-CoA
H_2O
H_2O

(3.5)

Fig. 3.1.

this system. TAL has been detected from several polyketide synthases incubated without NADPH [6], including fatty acid synthases [7,8] (see section 1.1.2), and appears to be a particularly stable 'shunt' or derailment product which forms on polyketide synthases when β-keto reductase activity is absent.

For the 6-MSA system, persuasive support for the timing of the single reduction at the C_6 and not the C_8 stage was obtained from inhibition studies on the synthase enzyme from *P. patulum* [9]. For example, the acetylene 3-pentynoyl-*N*-acetylcysteamine (3.12) completely inhibited 6-MSA production at low concentrations (10^{-5}–10^{-4} M). The initial rate of NADPH oxidation was still significant (60%), however, suggesting that the timing of inhibition occurred *after* the reduction event. The dehydratase site of 6-MSA synthase is believed to have a strategically located histidine residue which acts as a base to mediate an α,β-elimination of H_2O and generate the *cis* double bond. When presented with the acetylene (3.12), allene formation is promoted. The allene is then attacked by a histidine nitrogen, resulting in covalent attachment and irreversible inhibition.

Experiments have been designed to probe the stereochemistry of events operating during 6-MSA biosynthesis. Specific stereochemical insights have been obtained by a series of investigations utilizing chiral malonyl-CoA precursors. In separate experiments (2*R*) and (2*S*)[1-^{13}C, 2-^{2}H]-malonyl-CoAs, generated *in situ*, were incubated with acetyl-CoA, NADPH and 6-MSA synthase from *Penicillium patulum*. It was shown by comparison of the resultant mass spectra that the hydrogen atoms at C-3 and C-5 of 6-MSA arose from *different* prochiral hydrogens of malonyl-CoA, depending on whether deuterium had or had not been incorporated into 6-MSA. The conclusions drawn are consistent with a single base sitting above the folded C_6-hydroxy intermediate (3.15) which can mediate both the α,β-elimination and the enolization processes. This is illustrated in Fig. 3.2 for (2*S*)-methylmalonyl-CoA (3.13). The absolute stereochemistry of these events has also been established after extension of this experiment by incubating the chiral malonyl-CoAs, in separate experiments, with acetoacetyl-CoA (3.14). In these investigations only C-3 of 6-MSA will retain any of the original hydrogen atoms of malonyl-CoA, C-5 now originating from acetoacetyl-CoA. Mass spectral analysis clearly indicated that the C-3 hydrogen of 6-MSA had its origin in the 2-*pro-R* site of malonyl-CoA. By implication from the first experiment, the C-5 hydrogen must originate from the 2-*pro-S* site of malonyl-CoA. It is clear from the high degree of stereochemical control that the elimination and enolization processes are under enzymatic control. The interested reader is directed to the original papers for a detailed analysis and discussion of these results [10,11].

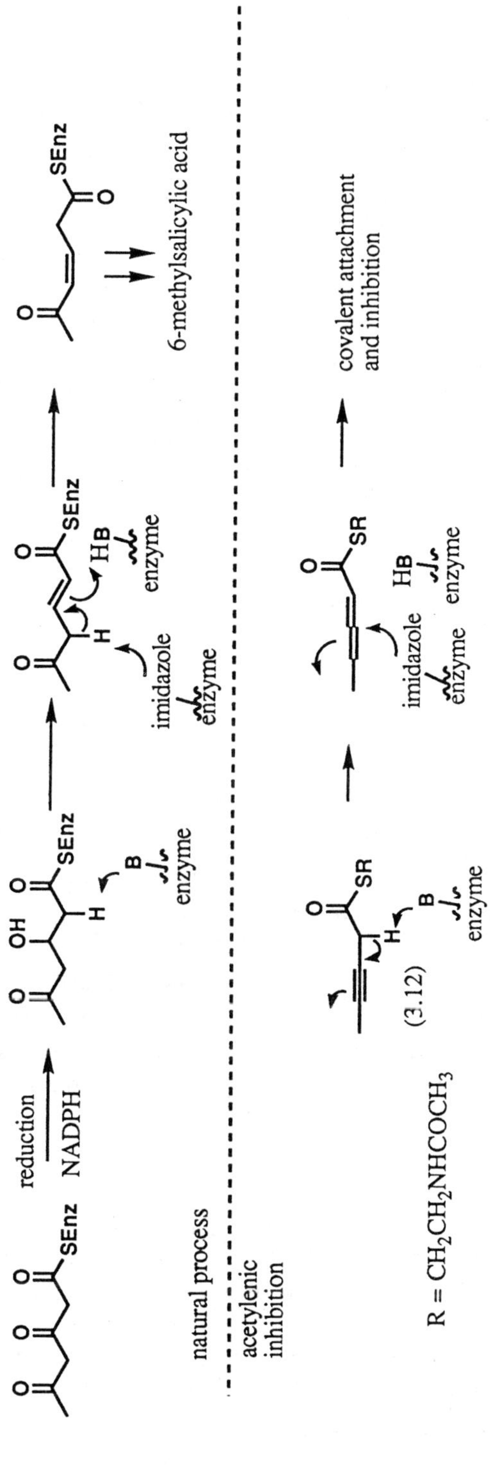
SEnz
reduction
NADPH
OH
H
B
enzyme
imidazole
enzyme
HB
6-methylsalicylic acid
natural process
acetylenic inhibition
SR
(3.12)
covalent attachment and inhibition
R = $CH_2CH_2NHCOCH_3$

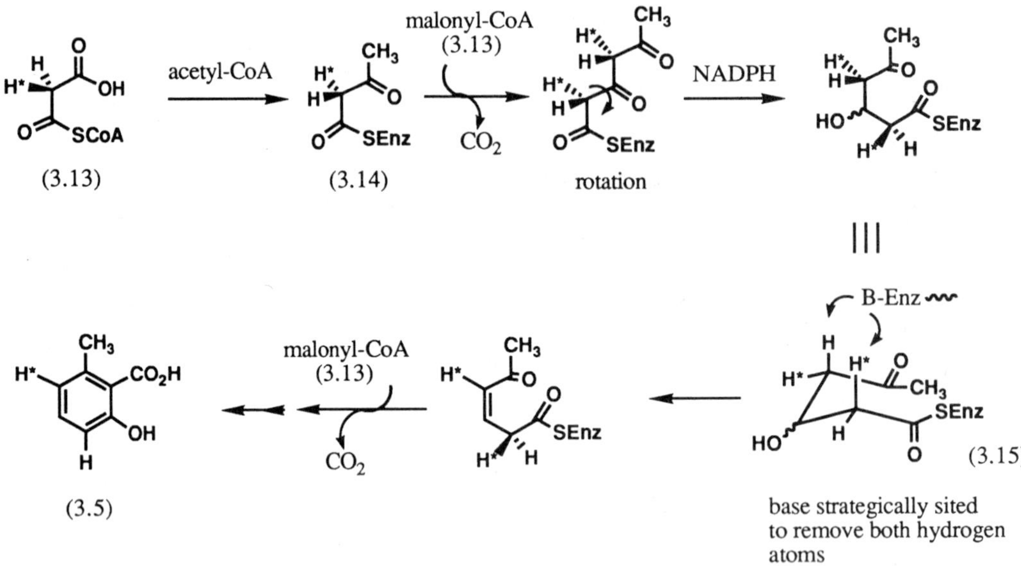

(3.13) (3.14) (3.15) (3.5)

Fig. 3.2.

3.1.2 Orsellinic acid

A similar series of investigations to those on 6-MSA (3.5), probing stereochemical control, have been carried out using orsellinic acid synthase from *Penicillium cyclopium* [12]. This synthase is operationally more straightforward than the 6-MSA synthase as there is no reduction/elimination required for the production of orsellinic acid. Incubation of (2*S*) and (2*R*)[1-^{13}C,^{2}H]-malonyl-CoAs revealed identical mass spectra from each experiment, characteristic of complete loss of stereochemical integrity during the transformation. After a most rigorous study and in complete contrast to 6-methylsalicylic acid, it was concluded that the enolizations occurring during orsellinic acid (3.3) biosynthesis are *non-stereospecific* and by implication are *not enzymatically* mediated.

3.1.3 A note on insects

Insects are not generally considered to be significant polyketide producers, but the limited available evidence does suggest that, in evolutionary terms, they have explored and exploited the polyketide pathways to a considerable extent. Some are known to synthesize prostaglandins (see section 2.2.2). Others can generate branched chain fatty acids derived from propionate (see section 1.3.1) and others large macrocyclic lactones from fatty acid metabolism [13]. More significant, however, are the aromatic polyketides such as methyl 6-methylsalicylic acid (3.16), mellein (3.17) and 2-hydroxy-6-methylacetophenone (3.18), which are apparently restricted to some ant species and are produced directly by polyketide synthases. These volatile aromatics are pheromones, excreted from the mandibular gland of the male ants, and often the same compound illicits a completely different response, e.g. alarm or swarming, in different species. Methyl 6-methylsalicylic acid (3.16) has been identified in the males of several ant species and is, for example, the main constituent of the pheromone cocktail of *Camponotus nearcticus* [14]. The biosynthesis of 2-hydroxy-6-methylacetophenone (3.18) has been studied in males of the Australian ponerine ant, *Rhytidoponera aciculata*. This investigation demonstrated an origin from acetate, and the classic folding of a pentaketide by one of two modes was proposed, with removal of the terminal carboxyl at some stage during the biosynthesis [15].

Although these compounds are typical of fungal and plant polyketides, their isolation under laboratory conditions and their occurrence only in the males of the species indicate that they are not of dietary or fungal origin. It is intriguing, and not obvious, how the intimate ecological interactions between plants and insects have resulted in the ability of these ants to synthesize such compounds.

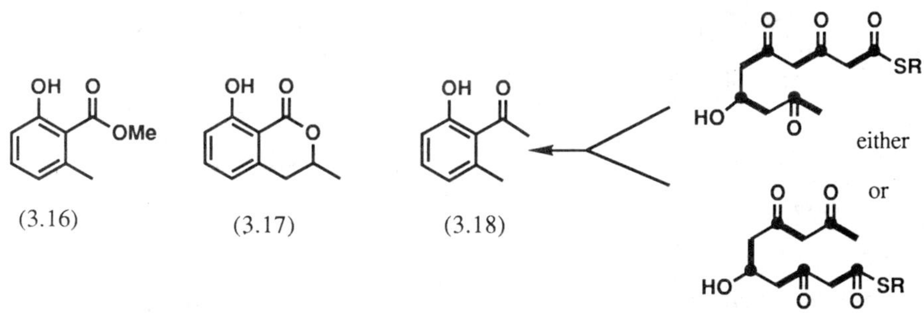

(3.16) (3.17) (3.18)

3.2 OTHER AROMATIC POLYKETIDES

3.2.1 Deuterium retention during aromatic ring formation

An interesting study on a variety of fungal metabolites set out to probe the nature of the enolization and reduction/elimination processes involved in assembling the aromatic ring systems, by comparing the relative ratios of deuterium retained from [$^2H_3$]-acetate at two specific sites, α and β, within each of the metabolites [16,17,18]. The construction of all the metabolites, (3.5) and (3.17)–(3.22), requires the folding of a polyketide chain and the generation of a *cis* double bond. The double bond is formed either by an enolization process when R=OH (as for (3.18)–(3.22)) or by reduction to an alcohol and then dehydration when R=H (as for (3.5) and (3.17)). The resultant data indicated that, during the enolization process, more deuterium is washed out at site β than at site α, whereas the reduction/elimination pathway gives rise to the opposite scenario, with a higher deuterium content at site β than at α. These observations are consistent with a reversible enolization process and an essentially irreversible reduction/elimination pathway.

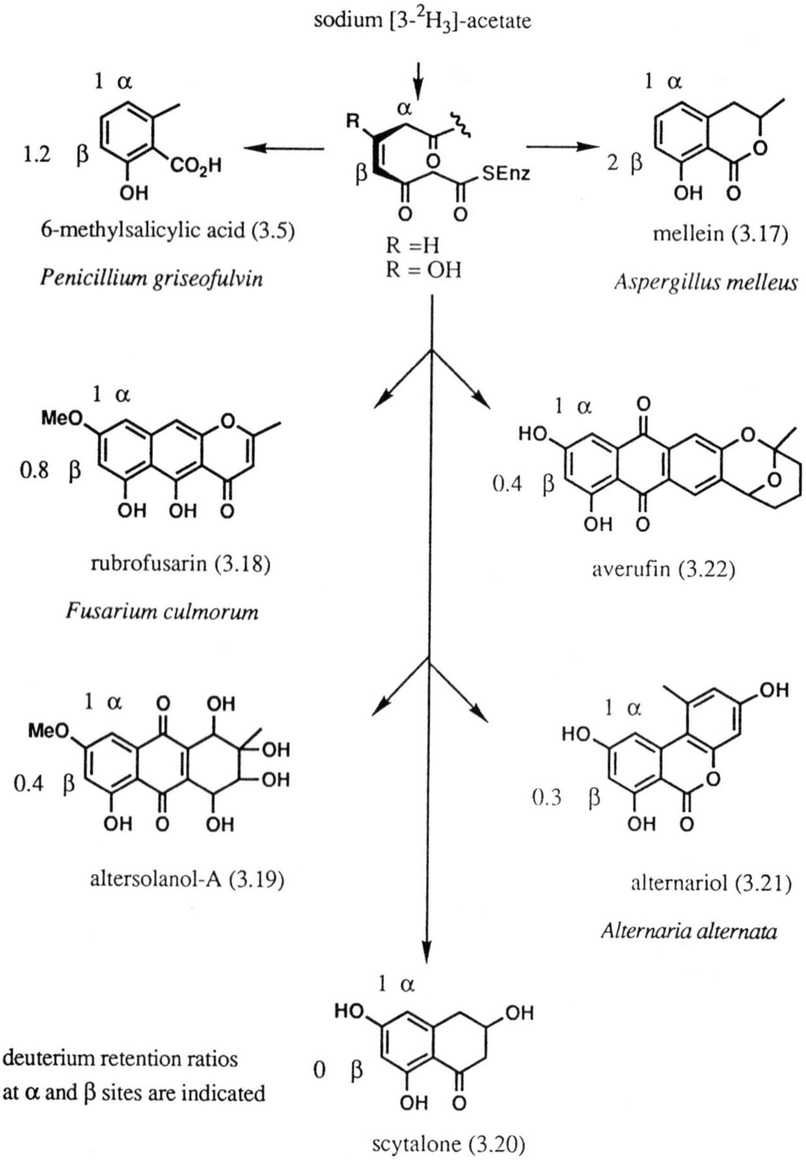

3.2.2 Naphthols and naphthaquinones

Naphthols and naphthaquinone metabolites occur to some extent in fungi and are extremely common in plants. In plants, however, this structural group is derived largely from shikimic acid and α-ketoglutarate rather than *via* a polyketide pathway (see section 3.2.3). Until relatively recently the origin of the fungal naphthalene derivatives was unclear, although a series of biosynthetic investigations appears to have established their polyketide origin, and indeed several different polyketide pathways have emerged in various fungal systems. Scytalone (3.20) and flaviolin (3.25) are co-metabolites of *Phialphora lagerbergii* and are derived from 1,3,6,8-tetrahydroxynaphthalene (3.24), a symmetrical precursor [19]. The intermediacy of (3.24) is implicated from [1,2-$^{13}C_2$]-acetate feeding experiments, where the intact bonds are scrambled equally between two orientations. For these compounds the carbon framework forms a cycle without a pendant methyl group and the identity of the starter unit is obscured. There is, however, circumstantial evidence to suggest that they are not derived from a pentaketide, but originate from a cyclized hexaketide such as the acylnaphthol (3.23), followed by deacylation. The incorporation of [$^{2}H_3$]-acetate into scytalone (3.20) revealed the retention of deuterium at C-4 and C-5, but not at C-2 and C-7, and this can be rationalized, at least to some extent, by deacylation of (3.23). A further experiment with ^{13}C-acetate displayed uniform enrichment of all the expected sites of scytalone (3.20) with no obvious 'starter' effect. The starter effect is characterized by one predominant enrichment due to

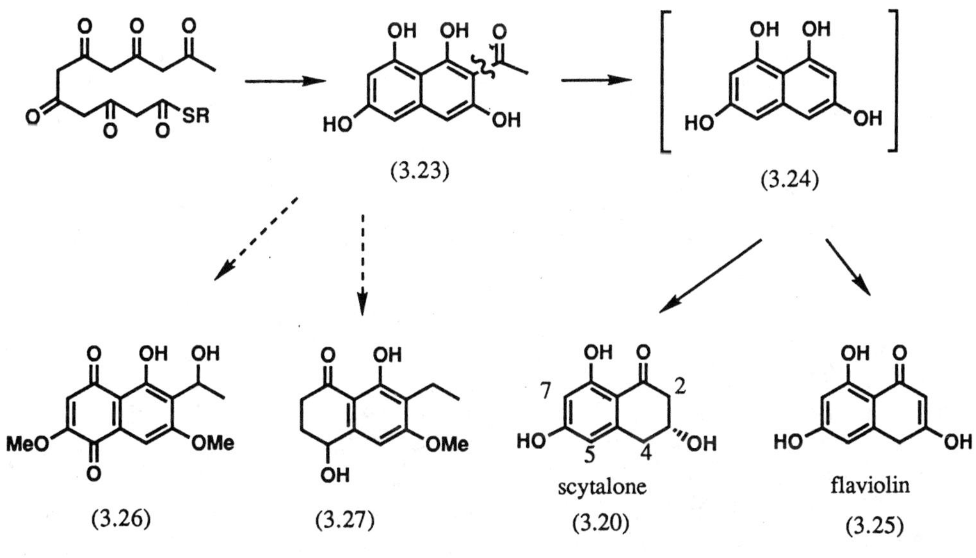

direct incorporation of acetyl-CoA from the acetate pool rather than indirect incorporation of the C_2 units via the malonyl pool. Other hexaketide metabolites such as (3.26) and (3.27) from *Scytaldium* sp. have a structural resemblance to scytalone, and could clearly derive too from the putative hexaketide (3.23).

Another and probably unique pathway for the generation of the naphthaquinone skeleton is exemplified by mollisin (3.28) [20]. It is a secondary metabolite of the fungus *Mollisia caesia* and contains an unusual dichloromethyl group. Mollisin may derive from two distinct polyketide chains where each of the terminal carboxyl groups are lost; however, an alternative and realistic scenario exists where a carbonyl has been carved from a loop of a continuous polyketide chain folding back on itself [21]. Finally, marticin (3.29), a naphthaquinone from *Fusarium martii*, provides a further pathway for the generation of this structural group. Marticin is of mixed biosynthetic origin and is derived from a hepataketide and a C_3 unit from the citric acid cycle [22].

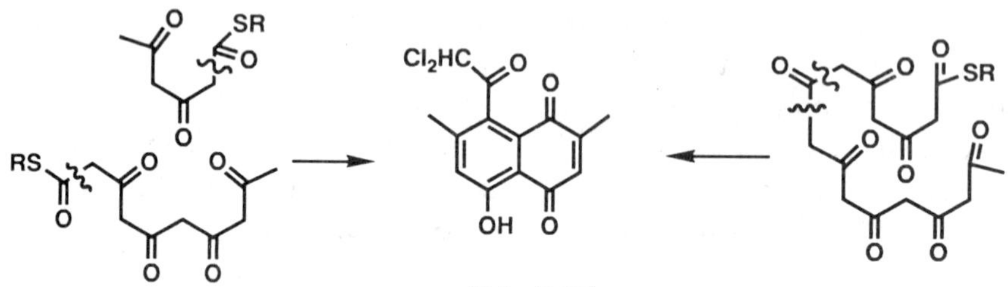

mollisin (3.28)

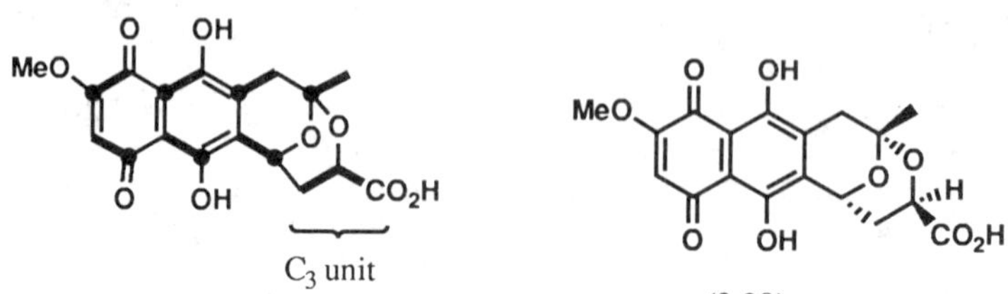

(3.29)

3.2.3 Anthraquinones

Anthraquinones are common fungal metabolites, and anthraquinones of both polyketide and shikimate origins occur in higher plants. The polyketide origin from acetate can be rationalized by the classic folding of a poly-β-keto chain. The origin of the fungal anthraquinone emodin (3.31) is a straightforward example arising from the intramolecular condensation of an octaketide, to presumably generate the anthracene (3.30), which if oxidized and decarboxylated gives rise to emodin (3.31). No reductions of the ketone functionality are required here.

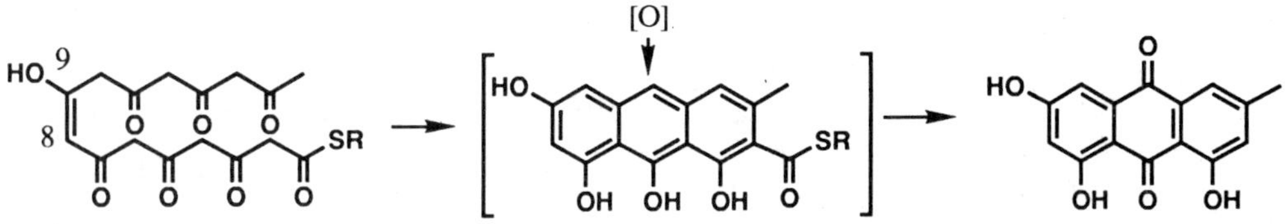

octaketide (3.30) emodin (3.31)

Aloesaponarin (3.32) is a plant anthraquinone derived from the folding of a polyketide chain; however, in higher plants an alternative and predominant pathway to anthraquinones involves shikimic (3.33) and α-ketoglutaric (3.34) acids and dimethylallyl pyrophosphate (DMAPP) derived from mevalonic (3.35) acid. The orange pigment alizarin (3.36) from *Rubia tinctorum L.* (madder plant) is a well-known example and its biosynthesis has been studied in detail [23]. Notably this leaves the A ring fully deoxygenated as a consequence of the biosynthetic pathway. There is no evidence to suggest such a pathway operating in fungal systems and alternatively, in fungi, anthraquinones fully deoxygenated on the A-ring can be generated directly from phenolic percursors. This has recently been established for the conversion of emodin (3.31) into chrysophanol (3.37) in *Pyrenochaeta terrestris*. An active cell-free extract from this system was found to require the co-factor NADPH [24]. A mechanism was proposed consistent with the incorporation of deuterium into C-6 of chrysophanol (3.37) from NADP^{2}H, and the exchange of deuterium into C-5 and C-7 when the deoxygenation was carried out in D_2O.

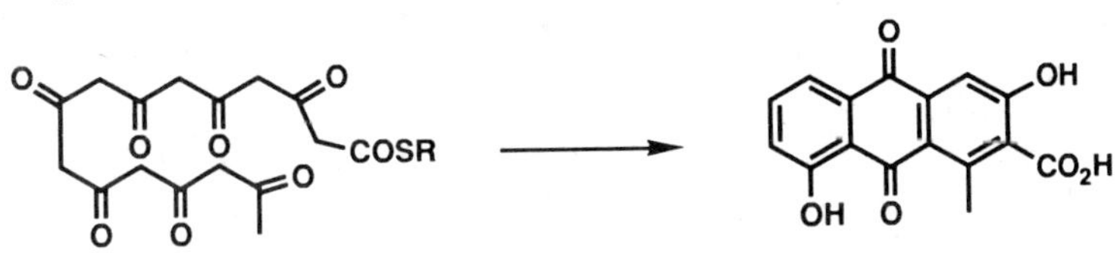

aloesaponarin (3.32)

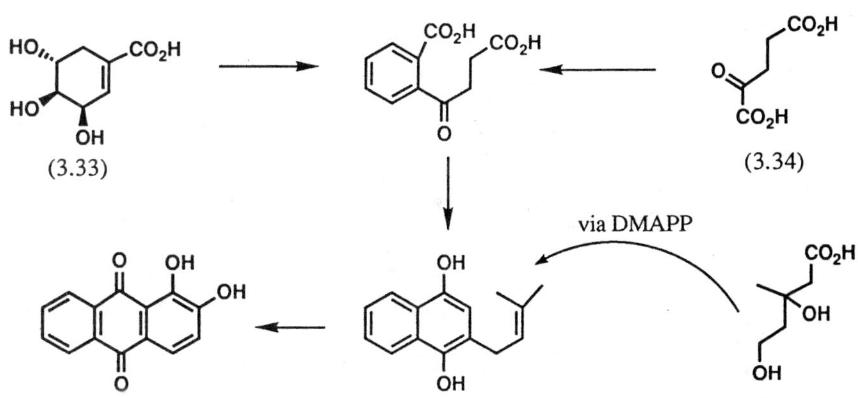

alizarin (3.36) (3.35)

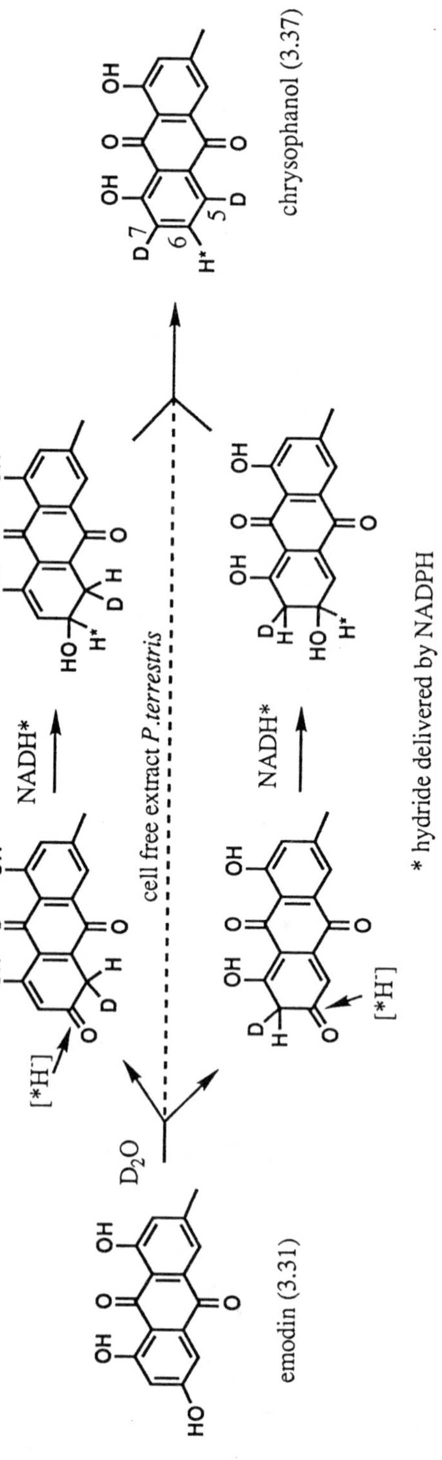
emodin (3.31)
D_2O
[*H⁻]
NADH*
cell free extract *P.terrestris*
NADH*
[*H⁻]
chrysophanol (3.37)
* hydride delivered by NADPH

3.2.4 Mechanism of anthraquinone formation

It is interesting to speculate at a mechanistic level on the various options which exist for the construction of the anthraquinone systems. In the first and conceptually most straightforward scenario, a polyketide is generated by the synthase and is stabilized on the protein surface in such an orientation that a series of simultaneous condensations clip the molecule together to form the anthracene. It has been suggested [25], however, that a more likely sequence of events would involve the initial formation of ring A to generate, for example, (3.38), followed by a cascade of cyclizations to form rings B and C in sequence. This 'zip fastener' hypothesis is attractive, as only one ring closure need be under enzymatic control, with other factors dictating the spontaneous formation of rings B and C. For the formation of ring B there are three possible cyclization modes, *a–a*, *b–b* and *c–c*. Predictably the *b–b* cyclization will be favoured over *a–a*, where the required anion necessitates abstraction of the least acidic protons, and *c–c*, where condensation would be to a sterically hindered ketone. Synthetic studies on model transformations lend some support to the zip fastener hypothesis [25].

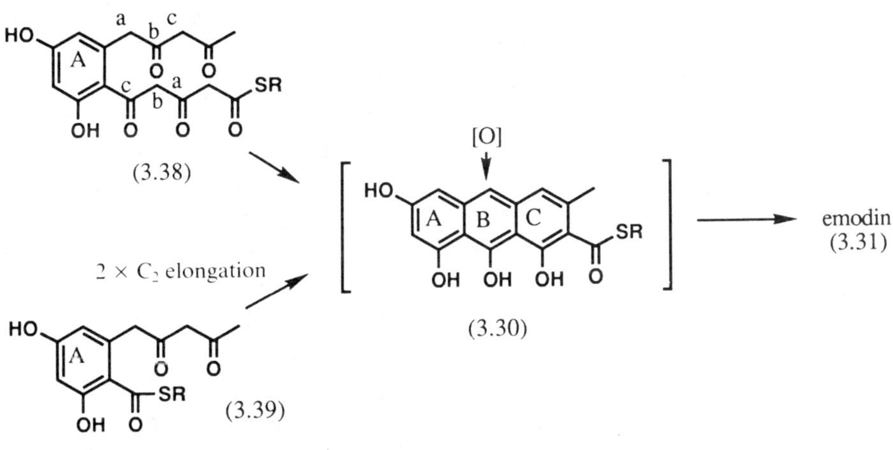

A processive pathway can also be considered where a hexaketide or some intermediate-sized polyketide chain such as (3.39) cyclizes to generate ring A and is then further extended by malonyl condensations, in a sequential manner, to provide ring B then C, followed by oxidation to the anthraquinone.

The intermediacy of an anthracene derivative such as (3.30) in emodin biosynthesis is consistent with all the biogenetic hypothesis presented above, although such compounds have not been identified in this role and their existence must remain speculative at present. Little is known too about the enzymes involved in introducing the second oxygen atom of the quinone moiety. An investigation on averufin (3.22) from a mutant strain of *Aspergil-*

lus parasiticus was shown to incorporate an atom derived from molecular oxygen at C-10 [26]; however, this is not surprising and does not reveal much detail on the overall process. It is entirely possible that the oxidation to the anthraquinones is a spontaneous, non-enzymatic transformation.

3.2.5 Xanthones

Xanthones of fungi and plants have different modes of biosynthesis. Fungal xanthones are in general acetogenic metabolites derived from the oxidative rearrangement of the anthraquinone skeleton. Ravenelin (3.43), from the lower fungi *Helminthosporium ravenelii* and *H. turcicum*, is a classic case and is believed to originate from the co-produced anthraquinones (3.40) and (3.41) by oxidative ring fission and loss of a carbonyl group. Subsequent cyclization generates the xanthone skeleton. Isotopic labelling studies utilizing [1,2-$^{13}C_2$]-acetate indicate that free rotation of an intermediate such as the benzophenone (3.42) must have occurred during the biosynthesis as the orientations of the intact bonds derived from acetate are scrambled [27].

The anthraquinone–xanthone pathway appears to be the most common route into fungal xanthones, although bikaverin (3.44), a metabolite of *Fusarium oxysporum*, offers an exception and is apparently derived directly from a folded polyketide chain. It could be argued, however, that bikaverin is closer to the naphthaquinones, with orsellinic acid (3.3) involved as a starter unit, although there is no supporting evidence for this at present [28].

Xanthones have a limited distribution in plants and are most commonly found in higher plants of the Guttiferae and Gentianaceae families [29]. The evidence suggests that the plant xanthones arise via the benzophenone (3.45) generated after three successive malonyl-CoA condensations with 3-hydroxybenzoyl-CoA. The majority of the plant xanthones can be rationalized as originating from this intermediate, and labelling studies have demonstrated the incorporation of (3.45) into selected xanthones from both of the above families [30,31]. Oxidative phenolic coupling provides the xanthone skeleton which is modified by various combinations of methylations, glycosylations and prenylations to generate individual xanthones. The timing of these events, and in particular the intramolecular phenolic coupling, is not established and could vary from one system to the next, therefore (3.46) may or may not be a common intermediate in xanthone biosynthesis. In a biosynthetic study on mangostin (3.47) from *Gentiana lutea*, cinnamic acid, benzoic acid, 3-hydroxybenzoic acid and (3.45) were all efficiently incorporated into mangostin, suggesting an origin of 3-hydroxybenzoyl-CoA from L-phenylalanine successively through these intermediates, most probably activated as their respective coenzyme-A esters [31]. These xanthones are more closely related to the flavonoid metabolites than to true acetogenic polyketides (see section 3.5.5).

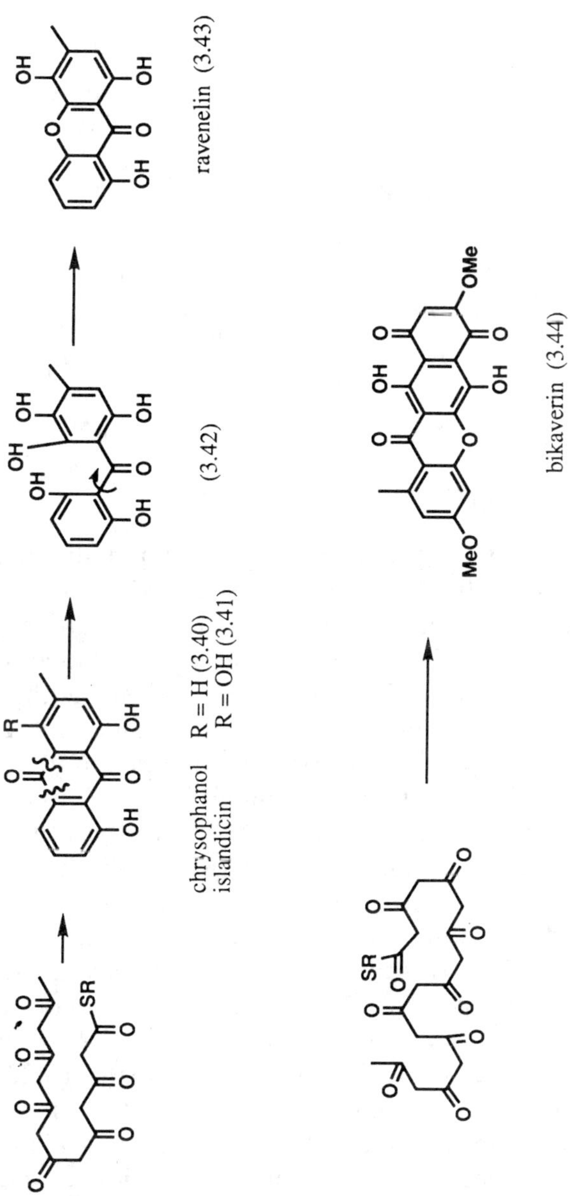
chrysophanol R = H (3.40)
islandicin R = OH (3.41)
(3.42)
ravenelin (3.43)
bikaverin (3.44)
SR
R
OH
OMe
MeO

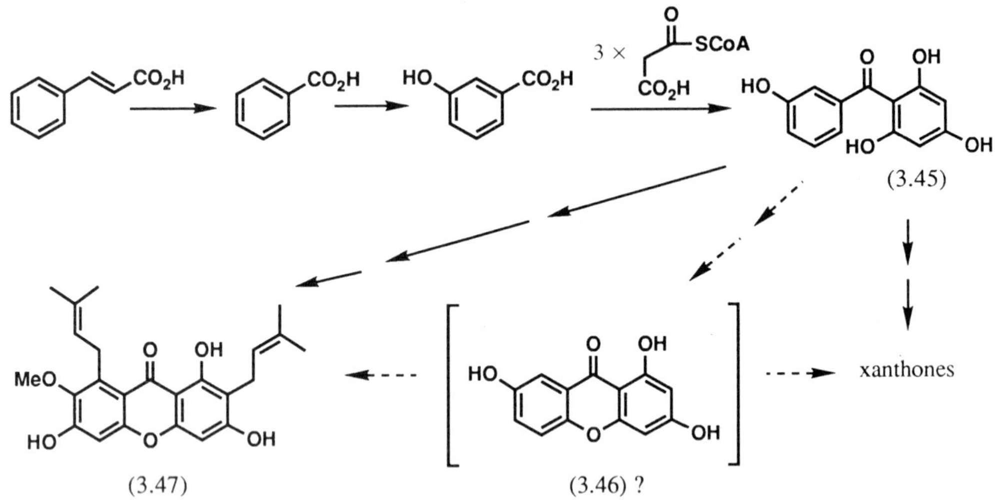

3.2.6 Aflatoxins

It is appropriate at this stage to include an account of aflatoxin biosynthesis. The aflatoxins are highly carcinogenic mycotoxins produced by several *Aspergillus* species and they have their origin in anthraquinones. The biosynthesis of aflatoxin B_1 (3.53) has received detailed scrutiny, and a complex pathway with some interesting features, including the rearrangement of the anthraquinone nucleus to a xanthone, has emerged.

Interestingly, hexanoate is incorporated intact into averufin (3.22), indicating a unique polyketide starter unit [32]. This C_6 initiator is apparent in the structures of norsolorinic acid (3.48) and averantin (3.49), the first identifable intermediates in aflatoxin-B_1 biosynthesis. It is noteworthy that an oxygen atom is *reintroduced* at C-5′ after averantin (3.49), a site originally accommodating the carbonyl oxygen for an acetate unit.

The anthraquinones, averufin (3.22), versiconal acetate (3.50) and versicolorin-A (3.51), have also been identified as intermediates after a series of mutant and inhibitor studies. Two blocked mutants of *Aspergillus parasiticus* have been selected, one of which accumulates averufin (3.22) and the other versicolorin-A (3.51). Addition of the insecticide dichlorvos (dimethyl-2,2-dichorovinyl phosphate) to the wild type strain inhibits aflatoxin-B_1 production, and versiconal acetate (3.50) becomes the major secondary metabolite [33]. Significantly sterigmatocystin (3.52), a carcinogenic mycotoxin produced by *Aspergillus versicolor*, is efficiently incorporated into aflatoxin-B_1 (3.53) and is therefore implicated as an advanced precursor on the pathway [34].

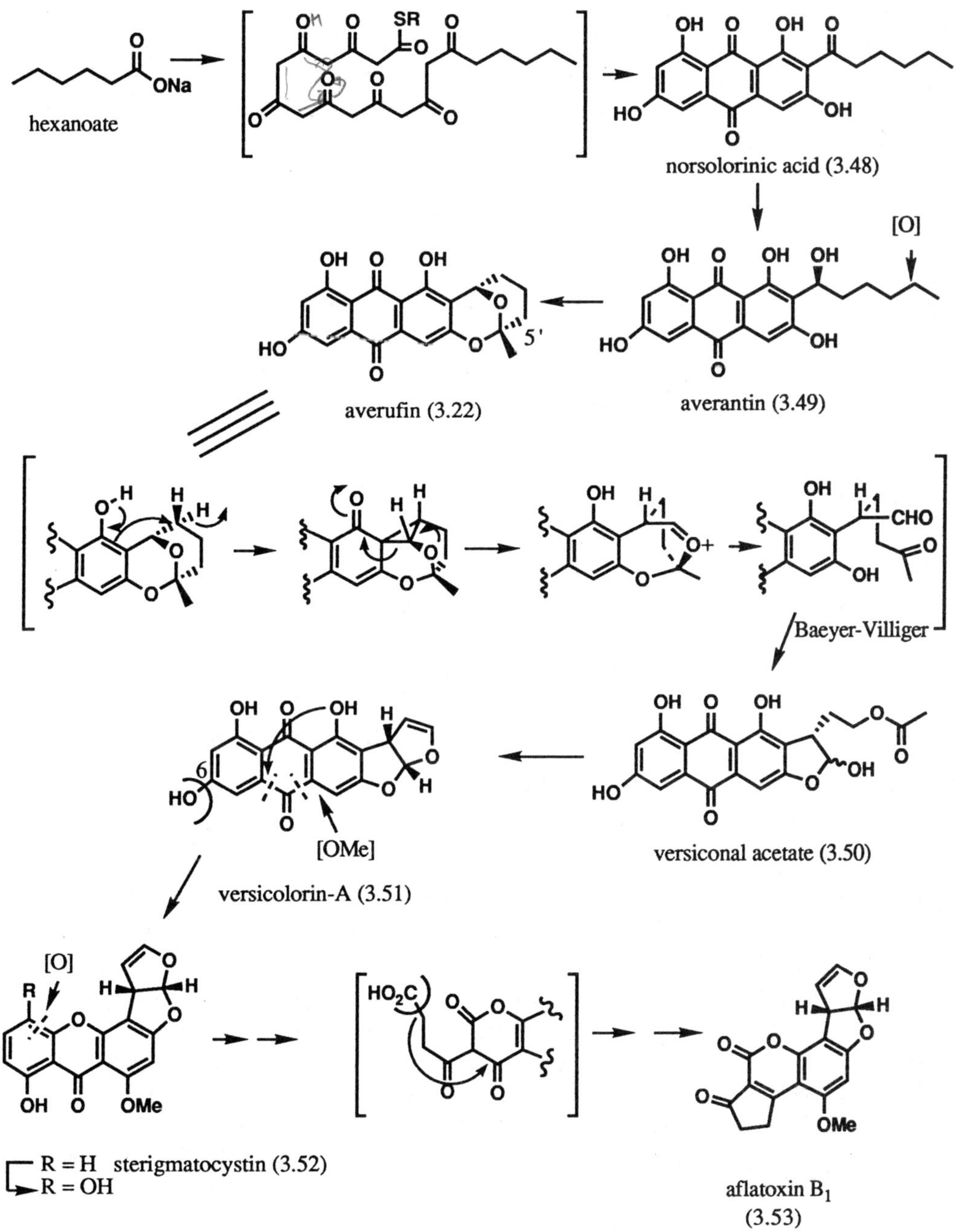
hexanoate
ONa
SR
norsolorinic acid (3.48)
[O]
averufin (3.22)
averantin (3.49)
5'
CHO
Baeyer-Villiger
versiconal acetate (3.50)
versicolorin-A (3.51)
[OMe]
[O]
OMe
HO_2C
R = H sterigmatocystin (3.52)
R = OH
aflatoxin B_1
(3.53)

This series of intermediates are firmly established in aflatoxin-B_1 biosynthesis and efforts are now focused on illuminating mechanistic details of the interconversions. For example the transformation of averufin (3.22) into versiconal acetate (3.50) requires a carbon skeletal rearrangement and the introduction of two oxygen atoms. The acetate carbonyl oxygen of (3.50) is derived intact from the C-5′-O of averufin (3.22) [35] and a mechanism involving cyclopropyl and oxonium intermediates, followed by hydrolysis and a Baeyer–Villiger-type oxidation, has been proposed consistent with experimental observations [36]. The conversion of the anthraquinone versicolorin (3.51) into the xanthone sterigmatocystin (3.52) involves substantial modification, and a satisfactory mechanism encompassing all the functional group conversions is not obvious. An oxidative ring fission and the subsequent loss of an original quinone carbonyl, presumably via a decarboxylation, is required. A methoxyl group is introduced (the methyl group and oxygen atom are not necessarily incorporated together) and an original phenolic alcohol swings round to become the xanthone ether oxygen. Perhaps the most remarkable aspect of the conversion of (3.51) into (3.52) is the reductive removal of the C-6 phenolic acid alcohol. The deoxygenation of a phenol has precedence in the conversion of emodin (3.31) into chrysophanol (3.37) in *Pyrenochaeta terrestris* (section 3.2.3) and clearly a similar transformation could occur during the conversion of versicolorin-A (3.51) into sterigmatocystin (3.52). Initial investigations probing such a process have shown that 6-deoxyversicolorin-A is not a precursor of aflatoxin-B_1 [37], although this does not rule out an NADPH-dependent deoxygenation on another versicolorin-A derivative.

3.3 EXTRA METHYL GROUPS

Many polyketide metabolites of fungi and plants possess 'extra' methyl groups which are inconsistent with a molecular skeleton derived from the folding of a polyacetate chain. These methyl groups are donated almost universally from the methyl group of L-methionine activated as *S*-adenosyl-L-methionine (SAM) (3.54). Some fungal metabolites have been isolated where, on initial analysis of the structure, it is tempting to suggest the involvement of propionate in the biosynthesis. Radiclonic acid (3.55) from a *Penicillium* sp. is such an example where there are five pendent, and two oxidized, methyl groups, which by structural analysis could be rationalized as originating largely from propionate. However, a biosynthetic investigation on this metabolite has demonstrated that *all* of the methyl groups derive from L-methionine [38]. The isolation and structure of cubensic acid (3.56) has been reported recently [39]. This is a metabolite from the higher fungus *Xylaria cubensis*, collected in the Amazon rain forest in Peru. An origin from acetate and propionate has been proposed, similar to the macrolide and polyether antibiotics of Streptomycetes (see section 4.4.1); however, caution must be exercised when making biosynthetic generalizations between different taxonomic groups. This would be unusual for a fungal metabolite and an origin from acetate, with the pendent methyl groups derived from L-methionine must remain the more likely possibility for cubensic acid biosynthesis. In a few cases in fungi, propionate is, however, utilized as a starter unit by some polyketide synthases which results in the introduction of an 'extra' methyl group at the beginning of the polyketide chain (see section 3.3.2).

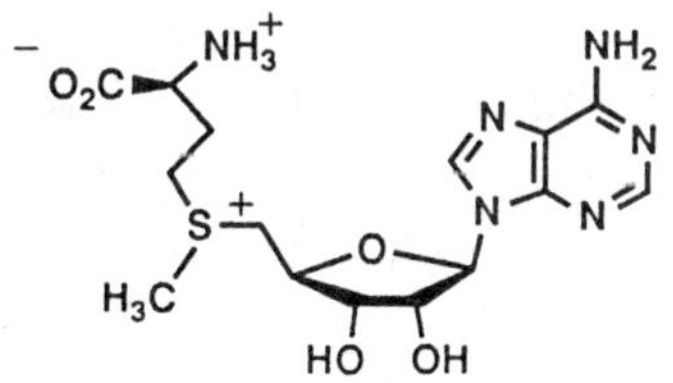

(3.54)

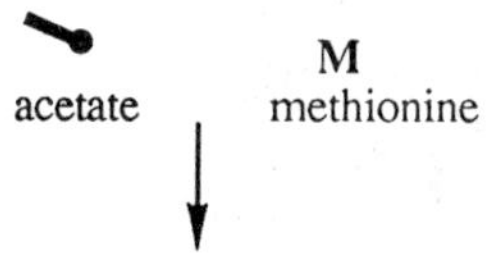

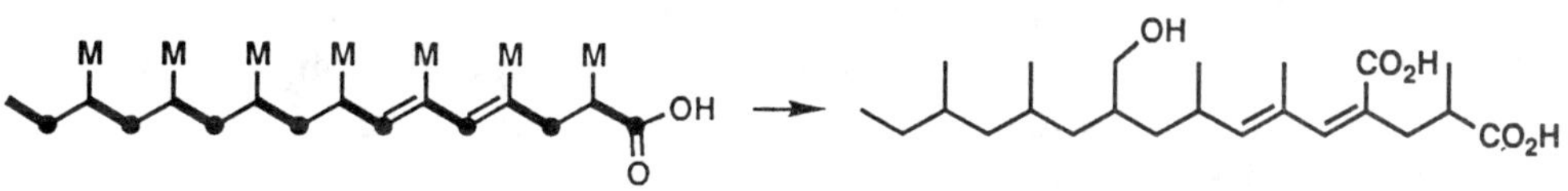

radiclonic acid (3.55)

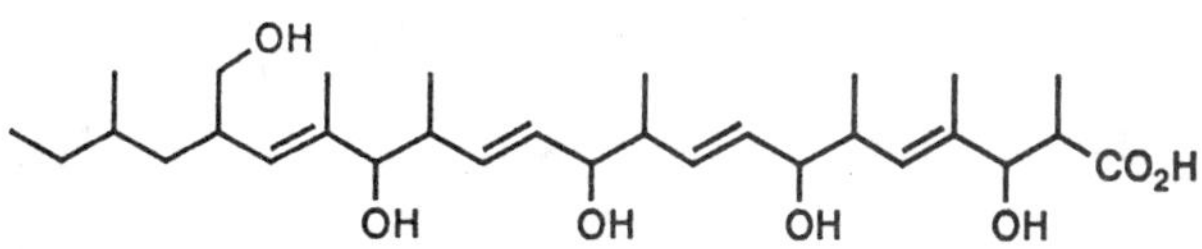

cubensic acid (3.56)

3.3.1 From *S*-adenosyl-L-methionine

The role of L-methionine as a methyl group donor was first demonstrated by Birch in 1958 during a biosynthetic investigation on mycophenolic acid (3.57). The study established the source of the methoxyl carbon atom and the nuclear C-5 methyl from the methyl group of L-methionine [40]. There are now numerous examples of such methylations in the literature and experiments employing [*methyl*-$^2H_3$]-methionine generally incorporate all three deuterium atoms into the metabolite. The stereochemistry of SAM methylations have been studied in many different instances, utilizing chiral [*methyl*-^1H,^2H,^3H]-L-methionine, and with few exceptions methylation proceeds by an inversion of configuration on carbon consistent with an S_N2 type process [41].

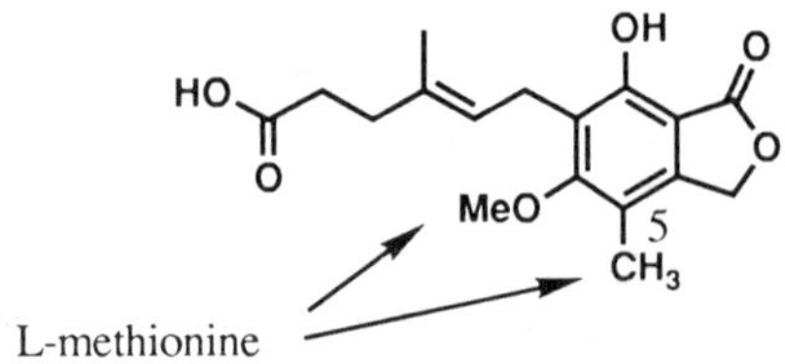

mycophenolic acid
(3.57)

The timing of C-methylations during polyketide biosynthesis is not well established, although it appears reasonable to conclude that they will occur at a nucleophilic centre such as a phenol or active methylene on a polyketide chain. Whether the methylating enzymes are integral with or separate from the polyketide synthase is an interesting question. For *O*- and *N*-methylations the latter appears to be the case as in most instances studied such modifications occur towards the end of biosynthetic pathways. *C*-methylations on the other hand probably take place much earlier, during polyketide assembly. This is almost certainly the case for reduced polyketides such as radiclonic acid (3.55) and sclerotiorin (3.58) which carry branched methyl groups derived from L-methionine. Methylation will not reasonably occur on a saturated chain and it is more probable that the methyl group of SAM is attacked either by a double bond or more likely by the active methylene of a poly-β-keto chain, prior to ketone reduction.

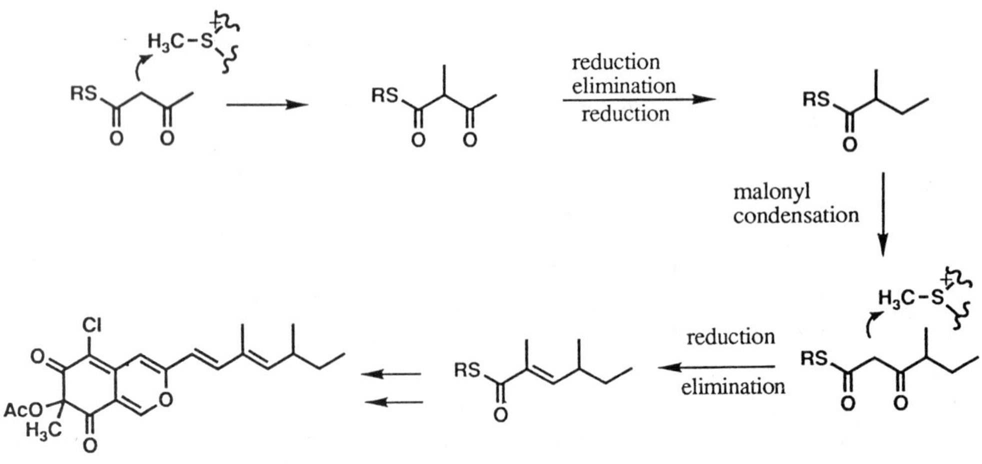

sclerotiorin (3.58)

3.3.2 From propionate

The origin of a methyl group from C-3 of propionate is unknown in plants and is rare in fungi. Recently, however, the anthraquinones austrocorticinic acid

(3.59) and (+)-austrocorticin (3.60) from the Australian toadstool *Cortinarius* sub sp. *dermocybe*, a higher fungus (Blasidomycota), were shown to have a propionate starter unit rather than the expected combination of acetate and a L-methionine-derived methyl group [42].

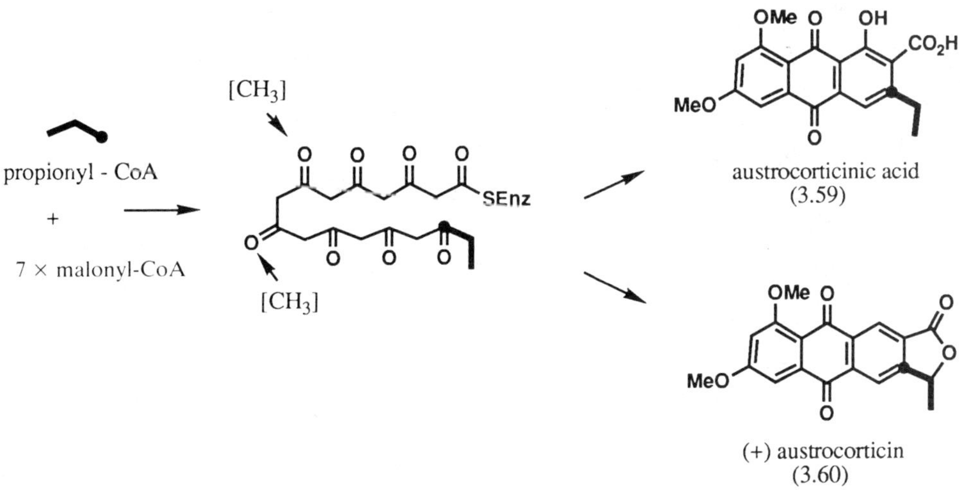

The structurally complex fungal polyketides aurovertins-B (3.61) and -D (3.62) and asteltoxin (3.63) are also remarkable as they are produced by two pathways which can apparently operate simultaneously in *Emericella variecolor* to generate the first three carbon atoms [43,44]. Amazingly, acetate, propionate and methionine are all incorporated into the C_3 starter unit. The first pathway involves a combination of acetate and the methyl group of L-methionine, whereas the second utilities propionate. The two can be unified conceptually if propionyl-CoA (3.64) is considered to be the true starter unit and the fungi generate methylmalonyl-CoA by methylation of malonyl-CoA as illustrated. This process is, however, speculative and remains to be verified.

Finally pseurotin-A (3.65) from the fungal strain *Pseudeurotium ovalis* is a metabolite of mixed biosynthetic origin. It is derived from the combination of L-phenylalanine and a polyketide chain, with an additional methyl group donated by L-methionine. The polyketide chain is constructed from four acetate subunits and a propionate starter unit [45]. Propionate incorporation into polyketides is common in bacteria, and polypropionate metabolites are increasingly being isolated from marine organisms (see, for example, section 5.4), but in these examples from fungi propionate is involved only as a chain initiator.

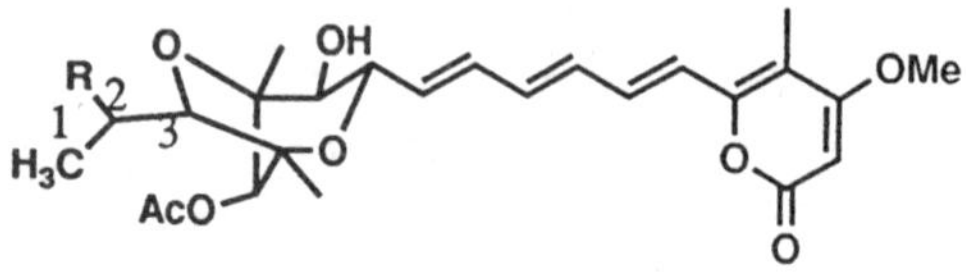

R = H aurovertin B (3.61)
R = OH aurovertin D (3.62)

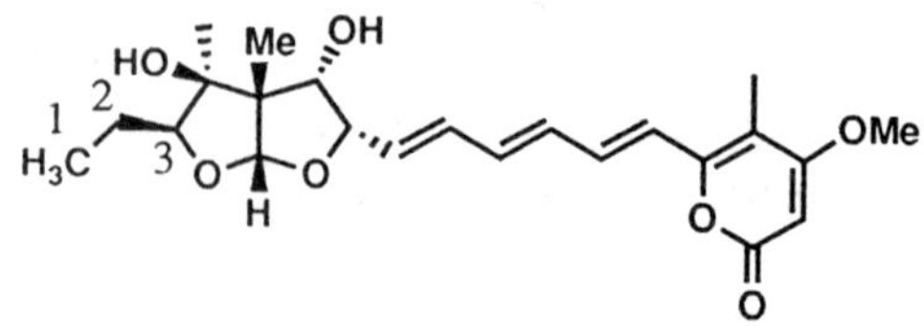

asteltoxin (3.63)

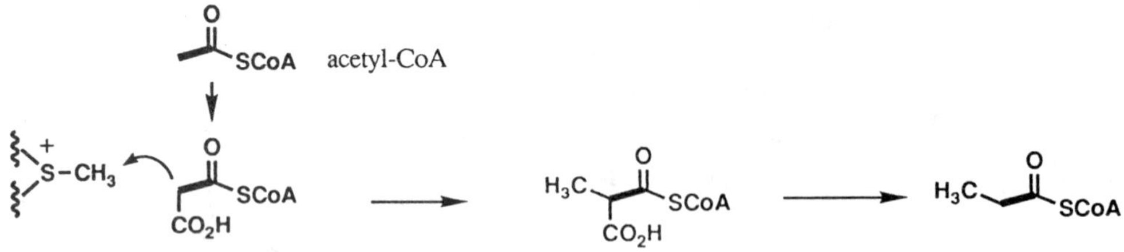

malonyl-CoA methylmalonyl-CoA propionyl-CoA (3.64)

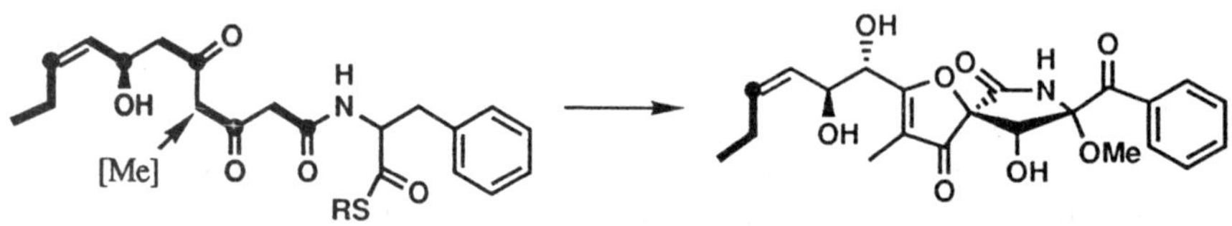

pseurotin A (3.65)

3.4 REDUCED FUNGAL POLYKETIDES

It has already been stated that the majority of the plant and fungal polyketides are aromatic in nature; however, a number of fungal metabolites have been isolated which have partially or wholly reduced polyketide frameworks. Biosynthetic investigations on these compounds are attractive, as a great deal of information can be gleaned from the increased level of processing that such metabolites receive by the producing synthases. The most significant mechanistic information emerging from these studies is the evidence that a *processive* mechanism is operating during the biosynthesis of the reduced polyketides (see section 3.4.3), a feature also established for the bacterial polyketides (see section 4.4.4). Stereochemical studies on these metabolites have revealed that some of the polyketide enzymes of secondary metabolism operate with the opposite stereoselectivity to those of primary metabolism (see section 3.4.4). Many of these partially reduced compounds are macrocyclic lactones and they often possess antibacterial activity, again a feature common to the bacterial polyketides. Several examples of these reduced metabolites are introduced.

3.4.1 Brefeldin-A and antibiotic A26771B

Brefeldin-A (3.67) is an antibacterial agent produced by several *Penicillium* species. It has structural characteristics similar to the prostaglandins, although it originates by a quite different biosynthetic pathway. It appears to be generated by a distinct synthase and not by modification of a C_{16} fatty acid. Labelling studies utilizing molecular oxygen and acetate carrying oxygen-18 have disclosed mechanistic details of brefeldin-A biosynthesis. The C-4 and C-7 hydroxyl oxygens are derived from molecular oxygen and a mechanistic rationale for the cyclopentane ring formation has emerged, consistent with an allylic rearrangement and epoxide ring opening of (3.66), concomitant with C–C bond formation [46].

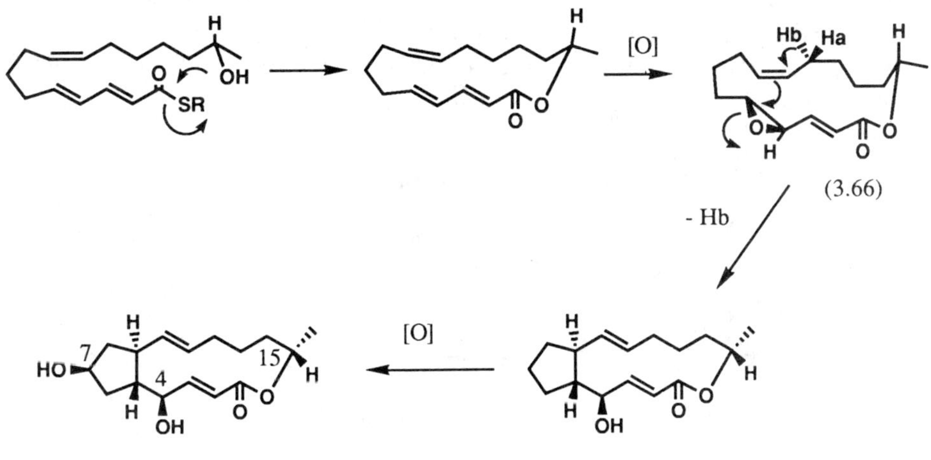

(3.66)

brefeldin-A (3.67)

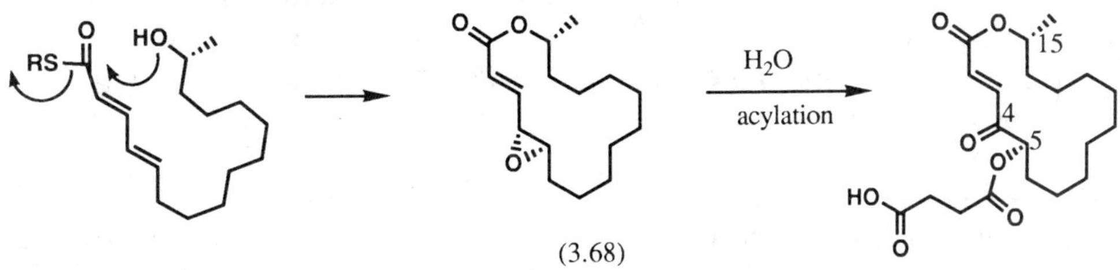

(3.68)

Antibiotic A26771B
(3.69)

Antibiotic A26771B (3.69) is a 16-membered macrocyclic lactone related structurally to brefeldin-A which also displays antibacterial activity. An analogous epoxide intermediate (3.68) can be invoked during its biosynthesis, which undergoes hydrolysis to a diol. Subsequent acylation at C-5 and oxidation to a ketone at C-4 generates (3.69). The origin of C-5-O from molecular oxygen and the C-4 keto oxygen from the aqueous medium is clearly consistent with this hypothesis [47]. Despite these close similarities it is noteworthy that brefeldin-A and antibiotic A26771B possess the opposite absolute configuration at C-15, indicating that their synthases possess β-keto reductase activities with opposite stereospecificities at this site.

3.4.2 Mevinolin and compactin

Mevinolin (3.70) and compactin (3.71) are metabolites of *Aspergillus terrus*. These compounds are powerful competitive inhibitors of 3-hydroxy-3-methylglutaryl coenzyme-A reductase (HMG-CoA reductase), the rate-limiting enzyme in steroid biosynthesis, and they have attracted a great deal of attention in recent years as agents for controlling *in vivo* cholesterol biosynthesis. Further, their emergence in the mid-1970s stimulated a tremendous synthetic activity in the pharmaceutical industry focused on designing simpler and more effective inhibitors such as (3.72) [48].

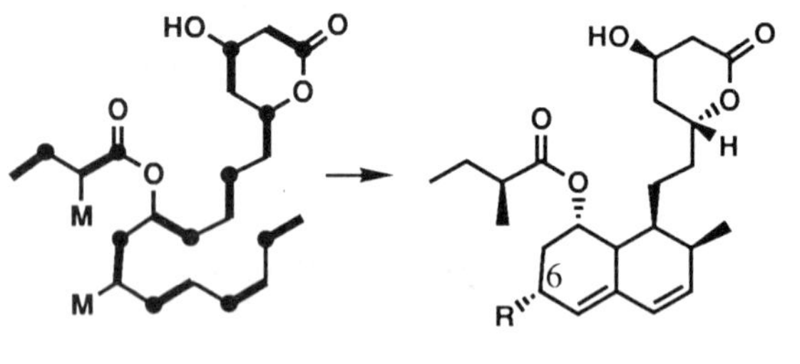

R = CH_3 (3.70)
R = H (3.71)

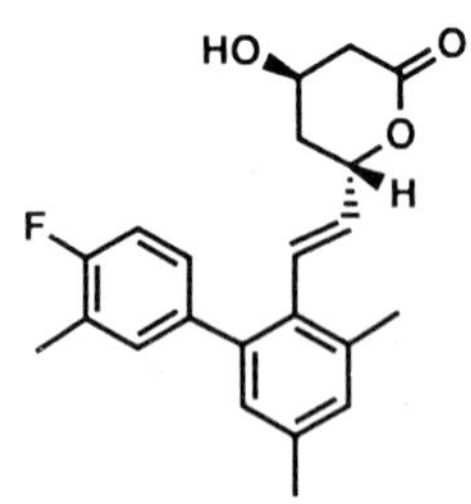

(3.72)

Mevinolin and compactin are derived from a reduced nonaketide, and the methyl group attached to C-6 of mevinolin (3.70) originates from L-methionine. Intriguingly, the 2-methylbutyryl side chain also originates from acetate and methionine rather than from isoleucine catabolism, perhaps the more obvious source for this fragment [49].

3.4.3 Dehydrocurvularin

Often in the literature, poly-β-keto-thioesters are drawn as precursors of polyketide metabolites. This is clearly valid for some simple aromatics, such as orsellinic acid, where no reductions are required and cyclization and enolizations afford the metabolite. In order to generate a reduced system from such a precursor the ketone functionality along the backbone of the poly-β-ketone requires adjustment to the correct reduction level. There is no evidence to support the direct reduction of a pre-assembled poly-β-keto-thioester. On the other hand, convincing evidence for a *processive* pathway operating in fungal polyketide biosynthesis comes from the intact incorporation of the thioesters (3.73) and (3.74) into dehydrocurvularin (3.75) in a mutant strain of *Alternaria cinerariae* [50]. The processive pathway suggests that the β-carbonyl generated after each malonate condensation is reduced to the requisite level *before* the next malonate condensation. The *N*-acetylcysteamine thioesters (3.73) and (3.74) apparently transesterify with (or replace?) the acyl carrier protein of the polyketide synthase, loading the C_6 and C_8 moieties onto the enzyme. Having intercepted the biosynthetic machinery, these fragments were further extended. The successful incorporation of dual ^{13}C labels from (3.73) and (3.74) into dehydrocurvularin (3.75) renders the direct reduction of a poly-β-keto-thioester, such as (3.76), an extremely unlikely process and places the processive mechanism on a firm footing. Additional cases of the incorporation of partially assembled fragments into polyketides have been successfully demonstrated in bacterial systems, raising similar mechanistic implications, and these are discussed in Chapter 4 (see section 4.4.4.).

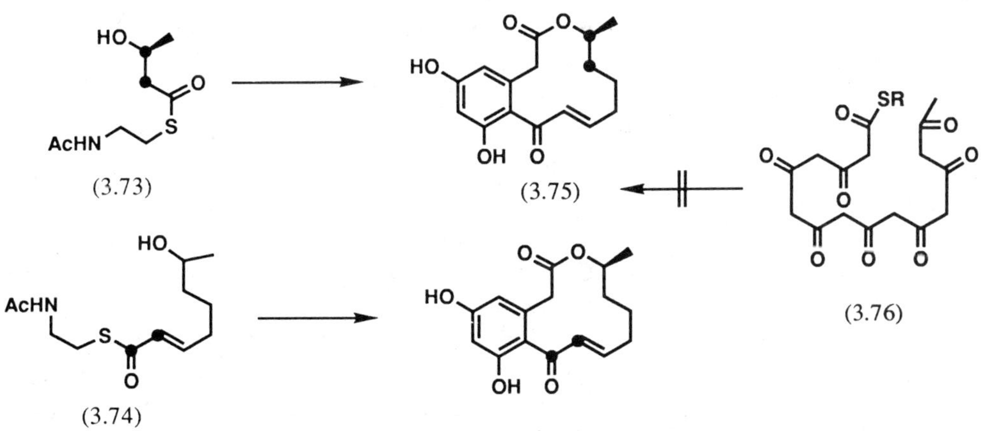

(3.73) (3.75) (3.76) (3.74) (3.75)

3.4.4 Stereochemistry of polyketide synthase enzymes

Biosynthetic investigations on the partially reduced fungal polyketides have not only provided valuable information on the assembly processes but have also revealed some stereochemical information on the synthase enzymes. In the discussion in Chapter 1 (see section 1.1.1) on the stereochemistry of the enzymes of fatty acid biosynthesis, it became apparent that all but one of the FAS enzymes had the same stereochemical course in all organisms. The exception was the last reducing enzyme, enoyl reductase, which displays variable stereoselectivity in different classes of organism. In fungi, the enoyl reductase enzymes of FASs all appear to display a uniform stereochemistry (see Table 1.2, page 25) and deliver hydrogen to the *si* face at C-2 of the enoyl thioester. It is interesting, therefore, to consider whether the stereochemistry of the polyketide synthase enzymes of secondary metabolism are uniform when compared with each other, and if they vary from their primary metabolite counterparts involved in fatty acid biosynthesis.

In most instances the β-keto reductase of fungal polyketide synthases generates the D-alcohol, similar to fatty acid biosynthesis; however, we have already seen (section 3.4.1) that the absolute stereochemistry at C-15 of brefeldin-A (3.67) and antibiotic A26771B (3.69) is opposite, indicating that the stereochemistry of the β-keto reductase enzymes generating these secondary metabolites is different and can vary from one system to the next. Indeed the two reduced centres at C-4 and C-15 of brefeldin-A (3.67) have the D- and L-configurations respectively, demonstrating that the *same* synthase can have similar activities operating with the *opposite* stereochemistry. This is not an unusual feature and is common also in the higher bacterial polyketides (see section 4.4.5).

In a particularly interesting survey [47,51,52,53] on the stereochemistry of *trans*-2-enoyl reductase, it has emerged that the enzyme of the fatty acid synthase and that of the polyketide synthase mediate reductions with the *opposite* stereochemistry in the *same* organisms! For example, after incubation of *Alternaria cinerariae* with [2-^{13}C,^{2}H$_3$]-acetate, the configuration of deuterium was assessed at the methylene sites derived from C-2 of acetate, in both the secondary metabolite dehydrocurvularin (3.75b) and oleic acid (3.77), produced simultaneously during the incubation. Analysis revealed that the isotope occupied the *pro-R* sites of oleic acid (primary metabolite) but occupied the *pro-S* sites at C-7 of dehydrocurvularin (3.75b) (secondary metabolite) [47]. A similar phenomenon was apparent after an analogous experiment with brefeldin-A (3.67b), where deuterium occupied the *pro-S* sites at C-6 and C-8 [53]. Only four fungal systems have been studied in this regard and although the sample is small the phenomenon would appear to be general. It is interesting too to note that averufin (3.22), an intermediate in aflatoxin biosynthesis, is the only case to emerge where deuterium labelled the *pro-R* site of a methylene carbon in a fungal polyketide. The starter unit during averufin biosynthesis is a C_6 hexanoate unit (see section 3.2.6), which may therefore have been constructed by a fatty acid synthase prior to priming averufin biosynthesis.

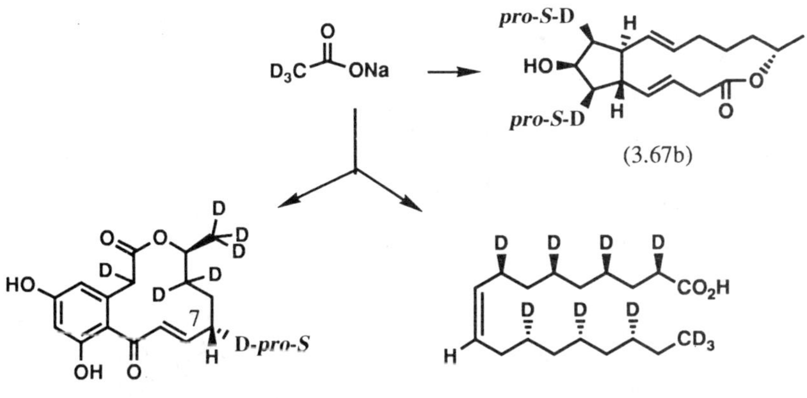

(3.67b)

(3.75b)

(3.77)

3.5 METABOLITES OF MIXED BIOSYNTHETIC ORIGIN

3.5.1 Terpene–polyketide metabolites

It is not uncommon to find polyketide metabolites in plants and fungi with pendent isoprenoid groups. These are generally attached to phenol ring systems, probably by an electrophilic aromatic substitution process on the corresponding isoprenoid pyrophosphates. The case studied in greatest detail is mycophenolic acid (3.57) from *Penicillium brevicompactum* [54]. 5-Methylorsellinic acid (3.78) is effectively incorporated into mycophenolic acid. Significantly, orsellinic acid itself is not a precursor, and this is consistent with methylation at C-5 occurring during polyketide assembly (see section 3.3.1). The intermediate lactone (3.79) is incorporated into (3.57) and the farnesyl-substituted derivative (3.80) has been isolated from the producing strain *P. brevicompactum*, supporting direct alkylation of the phenol nucleus.

Grifolin (3.81) is a straightforward example of a metabolite from the terpene–polyketide class derived from farnesyl pyrophosphate (C_{15}) and orcinol (3.6). We have seen, however, from mycophenolic acid (3.57) biosynthesis that the isoprenoid chain can undergo substantial modification, and a more elaborate example in this regard is austalide D (3.82), from *Aspergillus ustus*, which arises after a series of ring cyclizations and oxidations of (3.80), the same precursor to mycophenolic acid (3.57) [55].

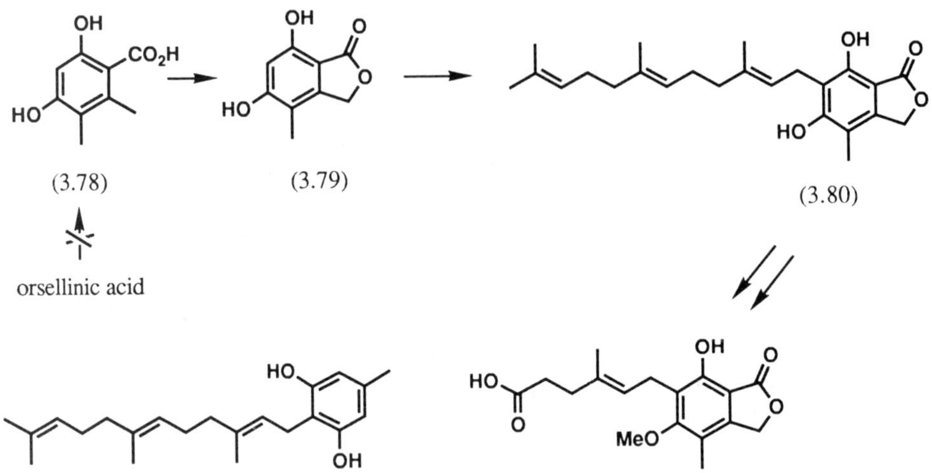

(3.78) (3.79) (3.80)

(3.81) (3.57)

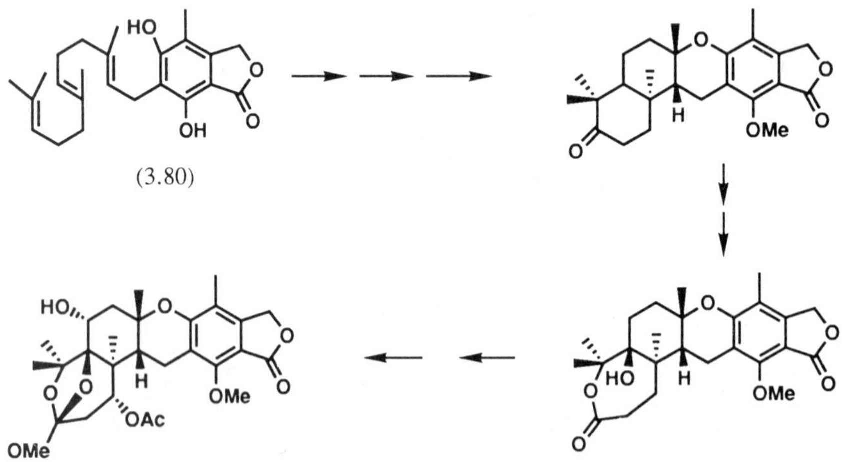

(3.80)

Austalide D (3.82)

3.5.2 Amino acid–polyketide

There are a variety of different structural classes of fungal metabolites which are derived from a combination of an amino acid and a polyketide moiety. Pseurotin-A (3.65) has already been highlighted as a polyketide–phenylalanine metabolite (see section 3.3.2). Tenellin (3.84), a yellow metabolite of *Beauveria bassiana*, and ilicicolin-H (3.85), from *Cylindrocladium ilicicola*, contain similar functionalized 2-pyridone ring systems and are derived from the condensation of a polyketide moiety with L-phenylalanine [56,57]. In each case the polyketide moiety has two L-methionine-derived methyl groups. L-Phenylalanine and not L-tyrosine is the more direct precursor, and a ring expansion of the acyltetramic acid quinone (3.83) has been proposed for the formation of tenellin (3.84). A similar process could account for ilicicolin-H (3.85). The cytochalasins, such as cytochalasin-B (3.86), are fungal antibiotics which also derive from a combination of L-phenylalanine and a polyketide chain, again containing two L-methionine-derived methyl groups [58]. Incorporation of molecular oxygen into the bridging oxygen of the lactone is consistent with a biological Baeyer–Villager-type oxidation as the final transformation in cytochalasin biosynthesis [59]. L-Phenylalanine appears to be the most common amino acid in this regard, although there are cases where other amino acids are involved. For example, fuligorubin-A (3.87), the yellow pigment of the slime mould *Fuligo septica*, is an acyltetramic acid metabolite derived from the condensation of glutamic acid and a polyene chromophore, which almost certainly has an acetogenic polyketide origin [60].

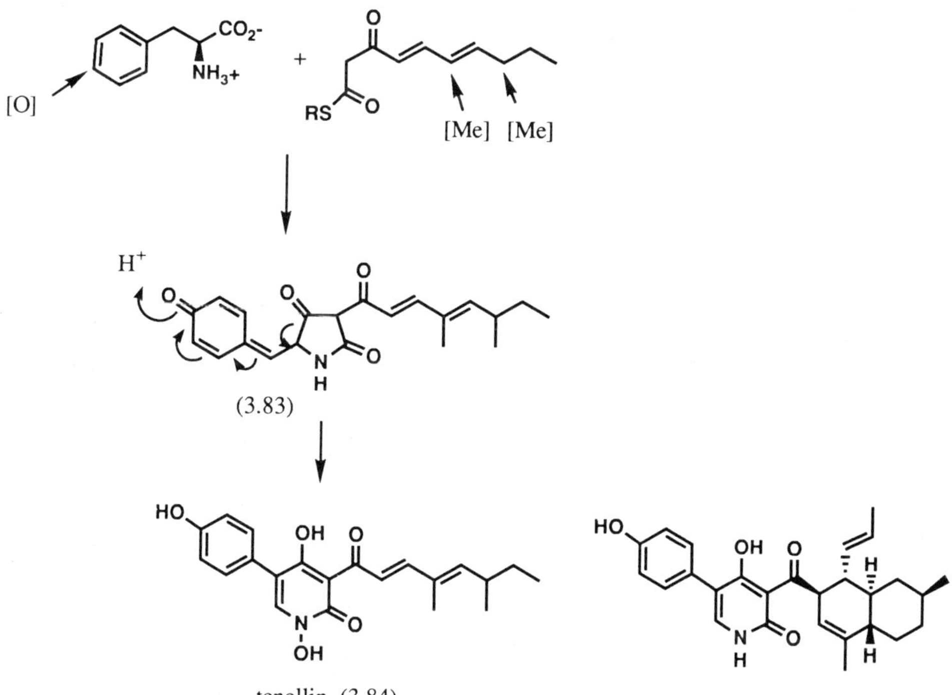

(3.83)

tenellin (3.84)

ilicicolin-H (3.85)

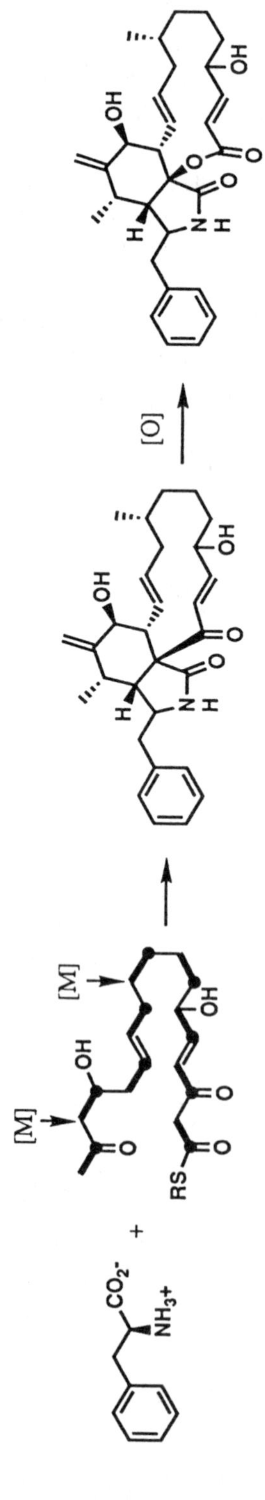

cytochalasin B (3.86)

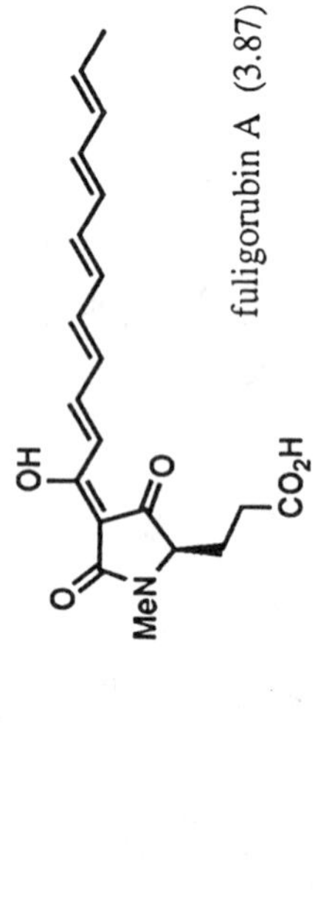

fuligorubin A (3.87)

3.5.3 Plant alkaloids

Plant alkaloids are in general biosynthesized from amino acids; however, a few notable exceptions exist which are polyketide in origin. The piperidine alkaloids coniine (3.91) (a hemlock alkaloid) and pinidine (3.92) are efficiently labelled by acetate. Coniine originates from the short chain saturated C_8 fatty acid octanoic acid. Oxidation occurs at C-5, and 5-oxo-octanoic acid (3.88), 5-oxo-octanal (3.89) and γ-coniceine (3.90) are efficiently incorporated into coniine (3.91) and are advanced precursors on the pathway [61]. Pinidine (3.92) is a C_9 alkaloid constructed from a pentaketide in an analogous manner to coniine. At some stage during biosynthesis the terminal carboxyl group is lost [62]. These alkaloids are however atypical and the piperidine alkaloids in plants generally derive from lysine.

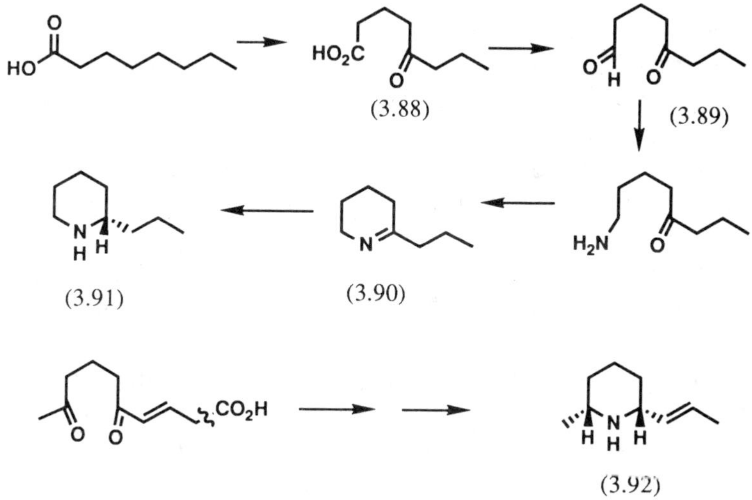

(3.88) (3.89) (3.90) (3.91) (3.92)

The acridine alkaloids originate after three malonate condensations on *N*-methylanthranilic acid (3.94). Anthranilic acid (3.93) has its origin in the shikimate pathway and, after *N*-methylation to (3.94), primes the biosynthesis of rutacridone (3.96) and arborinine (3.97). A cell-free system from *Ruta graveolens*, which produces rutacridone, was shown to convert *N*-methylanthranilic acid (3.94) and malonyl-CoA into 1,3-dihydroxy-*N*-methylacridone (3.95). The role of (3.95) as a precursor to rutacridone (3.96) is established, and this compound may also prove to be an intermediate on the arborinine (3.97) pathway [63].

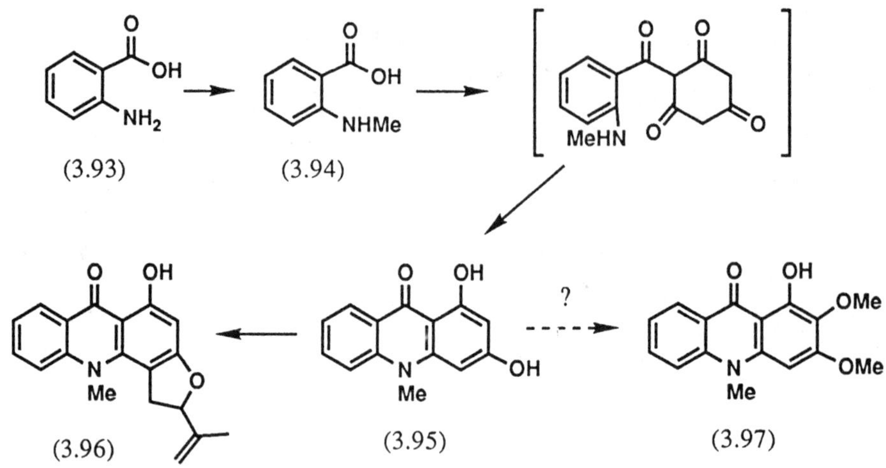

(3.93) (3.94) (3.96) (3.95) (3.97)

3.5.4 Flavonoids and related metabolites

The flavonoids and related compounds occur universally in plants. All derive biosynthetically from the condensation of three malonyl-CoA units onto a 4-coumaryl-CoA starter unit, a transformation mediated by the polyketide enzyme, chalcone synthase. A large number (more than 4000) [64] of structurally diverse metabolites have been characterized arising from this pathway, although the variety emerges from embroidering of the basic C_{15} chalcone skeleton (3.101) produced by chalcone synthase. Common modifications are *O*- and *C*-glycosylations, *O*-methylations and appendage of C_5 isoprene units, as well as oxidations to phenols and oligomerizations.

The ubiquity and structural diveristy of the flavonoids in the plant kingdom suggests that they have been the focus of tremendous selective pressures in evolution, and their functions are consequently varied and often subtle. Some clearly have a role to play in defence. Many display antifungal activity and are secreted by plants in response to fungal infections. The role of the pigmented anthocyanidins is at one level obvious, as they impart scarlet/blue colours to the petals of higher plants and are involved in the pigmentation of fruit; however, these compounds advantageously screen UV radiation and many display antifeedant activity against insects and larvae. Other flavonoids form high molecular weight polymeric substances, such as the tannins, which interact with proteins and polysaccharides providing a structural function as well as offering a medium for complex molecular communication and signalling between cells.

3.5.5 Flavonoid biosynthesis

The flavonoid pathway originates with L-phenylalanine. Phenylalanine ammonia lyase (PAL) deaminates the amino acid in a stereospecific manner to produce *trans*-cinnamic acid (3.98). This enzyme controls the flow of shikimate metabolites not only into flavonoid biosynthesis but also towards a great variety of phenylpropanoid secondary metabolites such as stilbenes, lignans and xanthones (see section 3.2.5). The next enzyme on the flavonoid pathway, cinnamate 4-hydroxylase, mediates *para* hydroxylation of *trans*-cinnamic acid to generate 4-coumaric acid (3.99). In order to prime the free carboxylic acid for condensation with malonyl-CoA, 4-coumaric acid (3.99) is transesterified to 4-coumaryl-CoA (3.100), the substrate for chalcone synthase.

Chalcone synthase controls the first committed step in flavonoid biosynthesis and has been purified from a number of sources and to homogeneity from parsley (*Petroselinum hortense*) [65]. It is a single protein which requires no co-factors, and *in vivo* it mediates the condensation of 4-coumaryl-CoA with three molecules of malonyl-CoA to generate tetrahydroxychalcone (3.101) as the product. Uniquely among polyketide condensing enzymes, chalcone synthase and the closely related enzyme, reservatrol synthase which generates reseveratrol, a precursor to the stilbenes phytoalexins as shown below, are devoid of an ACP (4′-phosphopantetheinyl moiety) and utilize the coenzyme-A thioesters of both substrates directly by transacylation onto specific cysteine thiols [66].

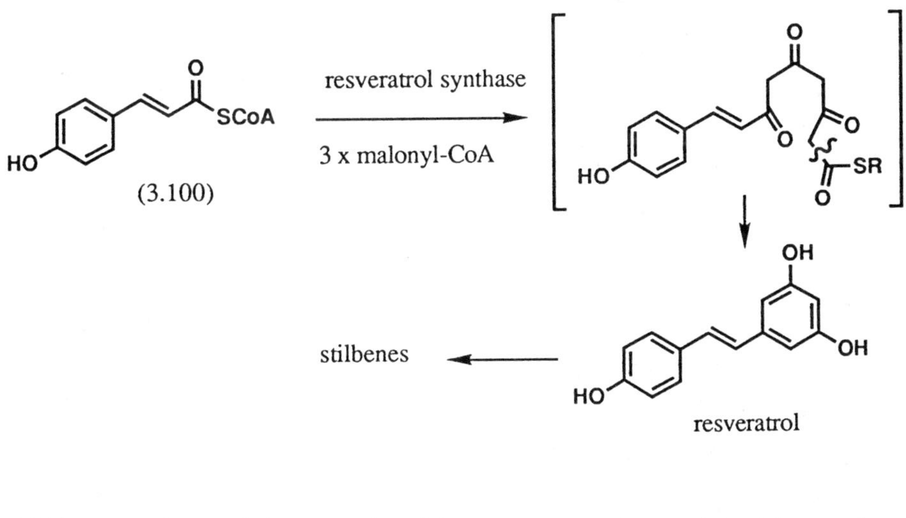

This is a substantial anomaly with respect to other polyketide synthases, although it is noteworthy that the *first* condensation in Type II fatty acid synthases employs acetyl-CoA directly (see section 1.1). *In vitro*, chalcone synthase will process a range of alternative starter units from butyryl-CoA to caffeyl-CoA, although their rates of conversion are in general poor [64]. Chalcones themselves are not widespread metabolites as they easily isomer-

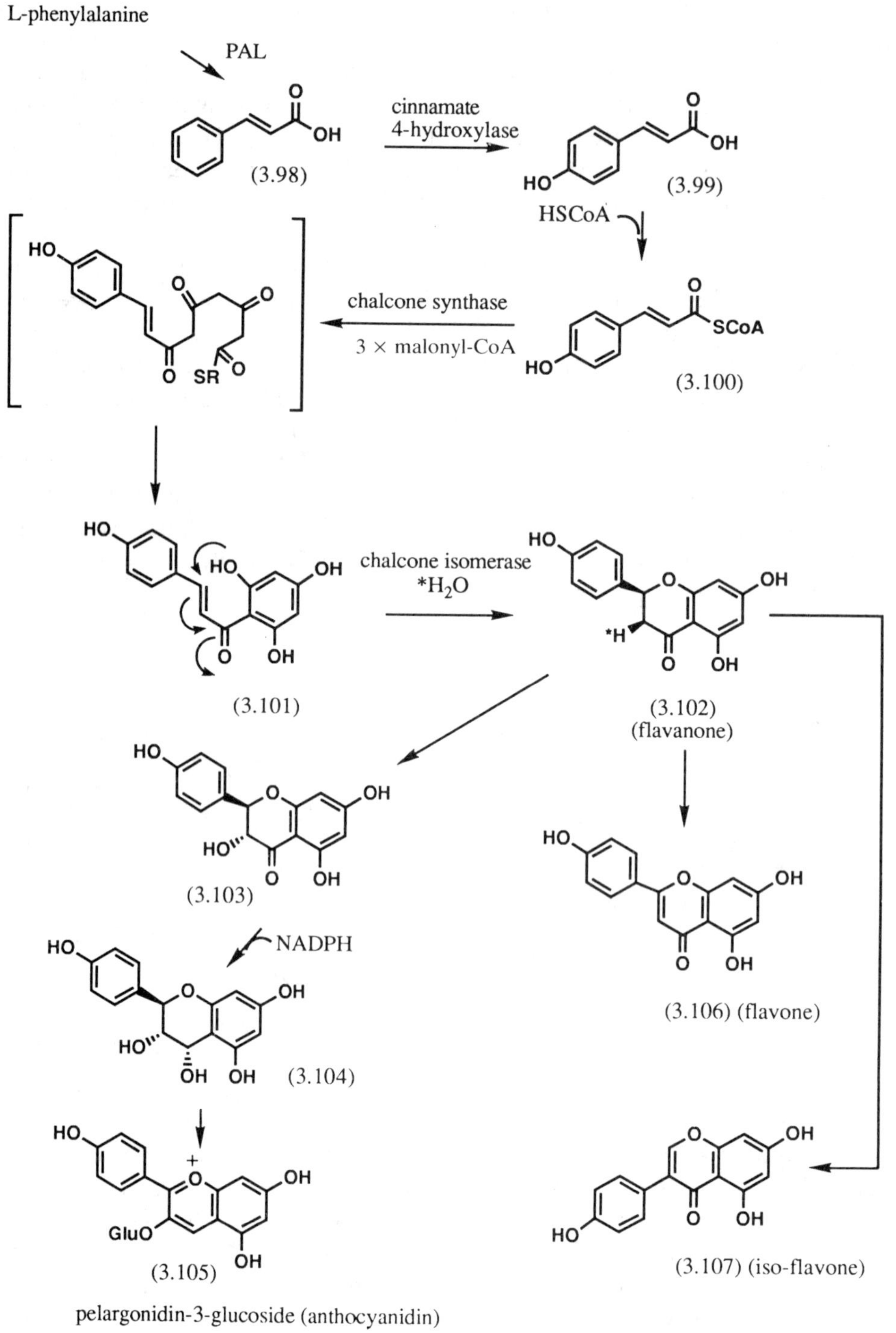
L-phenylalanine
PAL
(3.98)
cinnamate
4-hydroxylase
(3.99)
HSCoA
chalcone synthase
3 × malonyl-CoA
(3.100)
chalcone isomerase
*H_2O
(3.101)
(3.102)
(flavanone)
(3.103)
NADPH
(3.104)
(3.105)
pelargonidin-3-glucoside (anthocyanidin)
(3.106) (flavone)
(3.107) (iso-flavone)

ize to flavanones, although in Nature this is controlled by chalcone isomerase in a stereospecific manner, to give (2*S*)-flavanones such as naringenin (3.102). The steric course of this enzyme has been studies and addition to the double bond occurs in a *syn* mode [67].

The pathway from the flavanones to the anthocyanidin pigments involves hydroxylation at the 3-position. It has been demonstrated for example, that hydroxylation of naringenin (3.102) by flavanone-3-hydroxylase is specific for the (2*S*) enantiomer and generates (2*R*,3*R*)-dihydrokaempferol (3.103), a dihydroflavonol, as the sole product [68]. Proceeding towards the pyrylium ring system of the anthocyanidins, NADPH-dependent dehydrogenases reduce the ketone functionality of dihydroflavanols to generate flavan-3,4-*cis*-diols exemplified by (3.104). Such species aromatize readily under acidic conditions to form anthocyanidins, e.g. pelargonidin-3-glucoside (3.105); however, the nature of the enzymes controlling these steps is not clear.

In a branching pathway, modification of the flavanones to the flavones such as (3.106) requires the oxidative introduction of a double bond. The flavone synthases that catalyse this conversion all require NADPH but fall into two categories, i.e. class I or II, depending on whether they are Fe^{2+}/2-oxoglutarate- or cytochrome P_{450}-dependent [69]. Perhaps the most interesting transformation within the flavonoids is the conversion of a flavonoid to an isoflavonoid nucleus. The isoflavones do not have a wide taxonomic distribution, although they are common in the higher *Leguminosae*. Entry into these metabolites is typified by the rearrangement of naringenin (3.102) into genistein (3.107), involving a 2,3-aryl migration. The overall process is dependent on NADPH and molecular oxygen and recently a cytochrome P_{450} isomerase has been implicated, followed by a dehydratase activity [70]. A radical mechanism consistent with the involvement of cytochrome P_{450} is outlined in Fig. 3.4 (p. 104) which supersedes previous mechanistic proposals [70].

REFERENCES

[1] A. J. Birch and F. W. Donovan, Studies in relation to biosynthesis: some possible routes to derivatives of orcinol and phloroglucinol, *Aus. J. Chem.*, 1953, **6**, 360.

[2] A. J. Birch, R. A. Massey-Westropp, R. W. Rickards and H. Smith, Studies in relation to Biosynthesis. Part XIII. Griseofulvin, *J. Chem. Soc.*, 1958, 360.

[3] P. T. Izzo and C. R. Dawson, Cashew nut shell liquid VII. The higher olefinic components of cardanol, *J. Org. Chem.*, 1950, **15**, 707.

[4] L. K. Dalton and J. A. Lamberton, Studies of the optically active compounds of *Anacardiaceae* exudates: the long chain alicyclic keto-alcohol of tigaso oil, *Aus. J. Chem.*, 1958, **11**, 46.

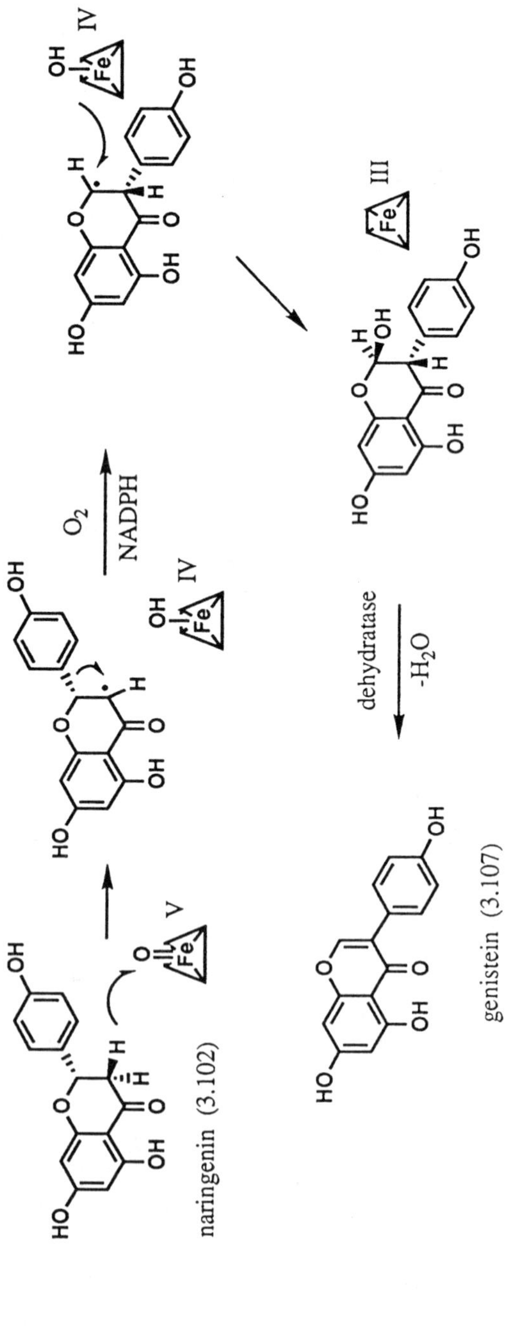
naringenin (3.102)
V
IV
O_2
NADPH
IV
III
dehydratase
$-H_2O$
genistein (3.107)

Fig. 3.4

[5] P. Dimroth, H. Walter and F. Lynen, Biosynthese von 6-methylsalicysaure, *Eur. J. Biochem.*, 1970, **13**, 98.
[6] R. Bentley and P. M. Zwitkowits, Biosynthesis of tropolones in *Penicillium stipitatum* VII. The formation of polyketide lactones and other nontropolone compounds as a result of ethionine inhibition, *J. Am. Chem. Soc.*, 1967, **89**, 676.
[7] G. E. Nixon, J. R. Putz and J. W. Porter, Synthesis of triacetic acid lactone by the pigeon liver fatty acid synthetase complex, *J. Biol. Chem.*, 1968, **243**, 5471.
[8] M. Yalpani, K. Willecke and F. Lynen, Triacetic acid lactone. A derailment product of fatty acid biosynthesis, *Eur. J. Biochem.*, 1969, **8**, 495.
[9] A. I. Scott, L. C. Beadling, N. H. Georgopapadakou and C. S. Subbarayan, Biosynthesis of polyketides. Purification and inhibition studies of 6-methylsalicylic acid synthase, *Bioorganic Chem.*, 1974, **3**, 238.
[10] P. M. Jordan and J. B. Spencer, Stereospecific manipulation of hydrogen atoms with opposite absolute orientations during the biosynthesis of the polyketide 6-methylsalicylic acid from chiral malonates in *Penicillium patulum*, *J. Chem. Soc. Chem. Commun.*, 1990, 238.
[11] J. B. Spencer, and P. M. Jordan, Use of chiral malonates to determine the absolute configuration of the hydrogen atoms eliminated during the formation of 6-methylsalicylic acid by 6-methylsalicylic acid synthase from *Penicillium patulum*, *J. Chem. Soc. Chem. Commun.*, 1990, 1704.
[12] E. R. Wo, I. Fujii, Y. Ebizuka, U. Sankawa, A. Kawaguchi, S. Huang, J. M. Beale, M. Shibuya, U. Mocek and H. G. Floss, Nonstereospecific proton removal in the enzymatic formation of orsellinic acid from chiral malonate, *J. Am. Chem. Soc.*, 1989, **111**, 5498.
[13] D. Vanderwel, H. D. Pierce Jr., A. C. Oehlschlager, J. H. Borden and A. M. Pierce, Macrolide (cucujolide) biosynthesis in the rusty grain beetle, *Cryptolestes ferrugineus* I', *Insect Biochem.*, 1990, **20**, 567.
[14] J. M. Brand, R. M. Duffield, J. G. MacConnell, M. S. Blum and H. M. Fales, Caste-specific compounds in male carpenter ants, *Science*, 1973, **179**, 388.
[15] B. Tecle, J. J. Brophy and R. F. Toia, Biosynthesis of 2-hydroxy-6-methylacetophenone in an Australian ponerine ant, *Rhytidoponera aciculata* (Smith), *Inset Biochem.*, 1986, **16**, 333.
[16] C. Abell, M. J. Garson, F. J. Leeper and J. Staunton, Biosynthesis of the fungal metabolites alternariol, mellein, rubrofusarin and 6-methylsalicylic acid from CD_3CO_2H, *J. Chem. Soc. Chem. Commun.*, 1982, 1011.
[17] C. Abell and J. Staunton, Biosynthesis of 6-methylsalicylic acid: the

combined use of mono and tri-deuterated acetate precursors to investigate the degree of stereochemical control in the aromatisation sequence, *J. Chem. Soc. Chem. Commun.*, 1984, 1005.

[18] T. J. Simpson, The biosynthesis of polyketides, *Nat. Prod. Rep.*, 1984, 281.

[19] E. Bardshiri and T. J. Simpson, ^{13}C and ^{2}H labelling on the biosynthesis of scytalone in *Phialaphora Lagerbergii*, *Tetrahedron*, 1983, **39**, 3539.

[20] L. W. Cary, H. Seto and M. Tanabe, Utilisation of ^{13}C-^{13}C coupling in structural and biosynthetic studies. The fourier transform ^{13}C NMR spectrum of mollisin, *J. Chem. Soc. Chem. Commun.*, 1973, 867.

[21] R. B. Herbert, *The biosynthesis of secondary metabolites*, London/ New York, Chapman and Hall, 2nd Edn., 1989, p. 50.

[22] J. E. Holenstein, A. Stoessl, H. Kern and J. B. Strothers, the marticins: conformation of structure, elucidation of biosynthetic origin by 13-C NMR studies, revision of stereochemical assignments, *Can. J. Chem.*, 1984, **62**, 1971.

[23] E. Leistner, Mode of incorporation of precursors into alizarin (1,2-dihydroxy-9,10-anthraquinone), *Phytochemistry*, 1973, **12**, 337.

[24] J. A. Anderson, B. K. Lin, H. J. Williams and A. I. Scott, Deoxygenation of phenolic natural products. Enzymatic conversion of emodin to chrysophanol, *J. Am. Chem. Soc.*, 1988, **110**, 1623.

[25] G. E. Evans, M. J. Garson, D. A. Griffin, F. J. Leeper and J. Staunton, Biomimetic synthesis of phenols from polyphenols, *CIBA Foundation Symposium*, 1978, **53**, 131.

[26] J. C. Vederas and T. T. Nakashima, Biosynthesis of averufin by *Aspergillus parasiticus*. Detection of ^{18}O-label by ^{13}C-NMR isotope shifts, *J. Chem. Soc. Chem. Commun.*, 1980, 193.

[27] A. J. Birch, J. Baldas, J. R. Hlubucek, T. J. Simpson and P. W. Westerman, Biosynthesis of the fungal xanthone ravenelin, *J. Chem. Soc. Perkin Trans. I.*, 1976, 898.

[28] A. G. McInnes, D. G. Smith, J. A. Walter, L. C. Vining and J. L. C. Wright, Homonuclear ^{13}C nuclear magnetic resonance studies of biosynthesis using doubly labelled precursors. Assembly pattern of the acetate units in bikaverin, *J. Chem. Soc. Chem. Commun.*, 1975, 66.

[29] G. J. Bennet and H.-H. Lee, Xanthones from Guttiferae, *Physiochemistry*, 1989, **28**, 967.

[30] P. Gupta and J. R. Lewis, Biogenesis of xanthones in *Gentiana lutea*. *J. Chem. Soc.* (C), 1971, 629.

[31] G. J. Bennet and H. H. Lee, The biosynthesis of mangostin: the origin of the xanthone skeleton. *J. Chem. Soc. Chem. Commun.*, 1988, 619.

[32] C. A. Townsend and S. B. Christensen, Stable isotope studies of anthraquinone intermediates in the aflatoxin pathway, *Tetrahedron*, 1983, **39**, 3575.

[33] R. Singh and D. P. H. Hsieh, Aflatoxin biosynthetic pathway: elucidation by using blocked mutants of *Aspergillus parasiticus*, *Arch. Biochem. Biophys.*, 1977, **178**, 285.

[34] D. P. H. Hsieh, M. T. Lin and R. C. Yao, Conversion of sterigmatocystin to aflatoxin-B_1 by *Aspergillus paraciticus*, *Biochem. Biophys. Res. Commun.*, 1973, **52**, 992.

[35] C. A. Townsend and S. B. Christensen, Concerning the role of nidurufin in aflatoxin biosynthesis, *J. Am. Chem. Soc.*, 1985, **107**, 270.

[36] C. A. Townsend, S. G. Davis, M. Korceda and B. Hulin, A cationic model of the chain branching step in aflatoxin biosynthesis, *J. Org. Chem.*, 1985, **50**, 5428.

[37] T. L. Graybill, K. Pal, S. M. McGuire, S. W. Brobst and C. A. Townsend, The timing of aromatic deoxygenation in aflatoxin biosynthesis, *J. Am. Chem. Soc.*, 1989, **111**, 8306.

[38] H. Seto, T. Sasaki and H. Yonehara, Studies on the biosynthesis of radiclonic acid. *Tetrahedron Lett.*, 1977, 4083.

[39] R. L. Edwards, D. J. Maitland and A. J. S. Whalley, Cubensic acid, a novel polysubstituted C_{22} fatty acid from the fungus *Xylaria cubensis* (Mont.) Fr. with substituents and substitution pattern similar to the macrolide antibotics. *J. Chem. Soc. Perkin Trans I.*, 1991, 1411.

[40] A. J. Birch, R. J. English, R. A. Massy-Westropp, M. Slaytor and H. Smith, Studies in relation to biosynthesis part XIV. The origin of the nuclear methyl groups in mycophenolic acid. *J. Chem. Soc.*, 1958, 365.

[41] H. G. Floss, Preparation, analysis and biochemical applications of chiral methyl groups, *Methods in Enzymology*, 1982, **87**, 126.

[42] M. Gill and A. Gimenez, (+)-Austrocorticin and austrocorticinic acid, the first anthroquinones derived from a propionate triggered octaketide, *J. Chem. Soc. Chem. Commun.*, 1988, 1360.

[43] P. S. Steyn, R. Vleggaar and P. L. Wessels, Biosynthesis of the aurovertins B and D. The role of methionine and propionate in the simultaneous operation of two independent biosynthetic pathways, *J. Chem. Soc. Perkin Trans. I*, 1981, 1298.

[44] P. S. Steyn and R. Vleggaar, Biosynthesis of asteltoxin by cultures of *Emericella variecolor*. The role of propionate in the biosynthesis and evidence for a 1,2-bond migration in the formation of the bistetrahydrofuran moiety, *J. Chem. Soc. Chem. Commun.*, 1984, 977.

[45] P. Mohr and Ch. Tamm, Biosynthesis of pseurotin-A, *Tetrahedron*, 1981, **37**, Supplement No 1, 201.

[46] C. T. Mabuni, L. Garlaschelli, R. A. Ellison and C. R. Hutchinson, Biosynthesis of the C_{15}-macrolide antibiotics. 1. Biochemical origin of four oxygen atoms in brefeldin-A, *J. Am. Chem. Soc.*, 1979, **101**, 707.

[47] K. Arai, B. J. Rawlings, Y. Yoshizawa and J. C. Vederas, Biosynthesis

of antibiotic A26771B by *Penicillium turbatum* and dehydrocurvularin by *Alternaria cinerariae*: comparison of stereochemistry of polyketide and fatty acid enoyl thiol ester reductases, *J. Am. Chem. Soc.*, 1989, **111**, 3391.

[48] G. E. Stokker, A. W. Alberts, P. S. Anderson, E. J. Cragoe, Jr., A. A. Deana, J. L. Gilfillan, J. Hirshfield, W. J. Holtz, W. F. Hoffman, J. W. Huff, T. J. Lee, F. C. Novello, J. D. Prugh, C. S. Rooney, R. L. Smith and A. K. Willard, 3-Hydroxy-3-methylglutaryl coenzyme-A reductase inhibitors 3. 7-(3,5-Disubstituted-[1,1′-biphenyl]-2-yl-3,5-dihydroxy-6-heptenoic acids and their lactone derivatives, *J. Med. Chem.*, **29**, 1986, 170.

[49] J. K. Chan, R. N. Moore, T. T. Nakashima and J. C. Vederas, Biosynthesis of mevinolin. Spectral assignment by double quantum coherence NMR after high carbon-13 incorporation, *J. Am. Chem. Soc.*, 1983, **105**, 3334.

[50] Y. Yashizawa, Z. Li, P. B. Reese and J. C. Vederas, Intact incorporation of acetic derived di- and tetraketides during biosynthesis of dehydrocurvularin, a macrolide phytotoxin from *Alternaria cinerariae*, *J. Am. Chem. Soc.*, 1990, **112**, 3212.

[51] C. A. Townsend, S. W. Brobst, S. E. Ramer and J. C. Vederas, Stereochemical features of enoyl thiol ester reductase in averufin and fatty acid biosynthesis in *Aspercillus parasiticus*, *J. Am. Chem. Soc.*, 1988, **110**, 318.

[52] B. J. Rawlings, P. B. Reese, S. E. Ramer and J. C. Vederas, Comparison of fatty acid and polyketide biosynthesis: stereochemistry of cladosporin and oleic acid formation in *Cladosporium cladosporioides*, *J. Am. Chem. Soc.*, 1989, **111**, 3382.

[53] C. R. Hutchinson, L. Shu-Wen, A. G. McInnes and J. A. Walter, Comparative biochemistry of fatty acid and macrolide antibiotic (brefeldin-A) formation in *Penicillium brefeldianum*, *Tetrahedron*, 1983, **39**, 3507.

[54] L. Canonica, W. Kroszczynski, B. M. Ranzi, B. Rindone, E. Stantaniello and C. Scolastico, Biosynthesis of mycophenolic acid, *J. Chem. Soc. Perkin Trans. I*, 1972, 2639.

[55] A. E. de Jesus, R. M. Horak, P. S. Steyn and R. Vleggaar, Biosynthesis of austalide K, a meroterpenoid mycotoxin from *Aspergillus ustus*, *J. Chem. Soc. Chem. Commun.*, 1983, 716.

[56] J. L. C. Wright, L. C. Vining, A. G. McInnes, D. G. Smith and J. A. Walter, use of ^{13}C in biosynthetic studies. The labelling pattern in tenellin enriched from isotope-labelled acetate, methionine and phenylalanine, *Can J. Biochem.*, 1977, **55**, 678.

[57] M. Tanabe and S. Urano, Biosynthetic studies with ^{13}C. The antifungal antibiotic ilicicolon-H, *Tetrahedron*, 1983, **39**, 3569.

[58] M. Binder, J.-R. Kiechel and Ch. Tamm, Zur biogenese des antibioticums phomin 1. Teil: Die grundbausteine, *Helv. Chim. Acta*, 1970, **53**, 1797.

[59] J.-L. Robert and Ch. Tamm, Biosynthesis of cytochalasins Part 5. The incorporation of deoxaphomin into cytochalasin-B, *Helv. Chim. Acta*, 1975, **58**, 2501.

[60] I. Casser, B. Steffan and W. Steglich, The chemistry of the plasmodial pigments of the slime mould *Fuligo septica* (Myxomycetes). *Angew. Chem. Int. Ed. Engl.*, 1987, **26**, 586.

[61] E. Leete and J. O. Olson, Biosynthesis and metabolism of the hemlock alkaloids, *J. Am. Chem. Soc.*, 1972, **94**, 5472.

[62] E. Leete, J. C. Lechleiter and R. A. Carver, Determination of the starter acetate unit in the biosynthesis of pinidine, *Tetrahedron Lett.*, 1975, 3779.

[63] A. Baumert, G. Schneider and D. Groger, Biosynthesis of acridone alkaloids. A cell-free system from *Ruta graveolens* cell suspension cultures, *Z. Naturforsh.*, 1986, **41c**, 187.

[64] *The flavonoids: advances in research since* 1980 J. B. Harborne, Chapman and Hall, London, 1988.

[65] F. Kreuzaler, H. Ragg, W. Heller, R. Tesch, I. Witt, D. Hammer and K. Hahlbrock, Flavanone synthase from *Petroselinum hortense*: molecular weight, subunit composition, size of messenger RNA, and absence of pantetheinyl residue, *Eur. J. Biochem.*, 1979, **99**, 89.

[66] T. Lanz, S. Tropf, F-J. Marner, J. Schroder and G. Schroder, The role of cysteines in polyketide synthases, *J. Biol. Chem.*, 1991, **266**, 9971.

[67] R. A. Dixon, P. M. Dey and I. M. Whitehead, Purification and properties of chalcone isomerase from cell suspension cultures of *Phaseolus vulgaris*, *Biochem. Biophysica Acta*, 1982, **715**, 25.

[68] L. Britsch and H. Grisebach, Purification and characterisation of (2*S*) flavanone 3-hydroxylase from *Petunia hybrida*, *Eur. J. Biochem.*, 1986, **156**, 569.

[69] P. M. Dewick, The biosynthesis of the shikimate metabolites, *Nat. Prod. Rep.*, 1989, **6**, 263.

[70] T. Hakamatsuka, M. F. Hashim, Y. Ebizuka and U. Sankawa, P-450 dependant oxidative rearrangement in isoflavone biosynthesis: Reconstruction of P-450 and NADPH: P-450 reductase, *Tetrahedron*, 1991, **47**, 5969.

4

Bacterial polyketides

Bacteria in general are not particularly adept at producing polyketide secondary metabolites, certainly when they are compared with the fungi, but a sense of disproportion emerges when the Actinomycetes and in particular the sub-class Streptomyces are considered. These are soil bacteria responsible for the synthesis of a great variety of polyketides extending from aromatic compounds, typical of the fungi and plants, to some of the most structurally complex secondary metabolites so far identified in Nature. They have attracted a great deal of attention as a source of antibiotics and the clinical successes of the tetracycline and the macrolide antibiotics have ensured the continuation of screening programmes and as a result the regular discovery of new metabolites and the elucidation of their structures.

The structural complexity of many of these metabolites arises from the utilization during chain extension not only of acetate but also of propionate and to some extent butyrate, resulting in the introduction of alkyl functionality along the polyketide backbone (see section 4.4.2). Further, the polyketide synthases can select to retain or reduce to various levels the β-ketone functionality which arises during polyketide assembly. The synthases possess the whole complement of reducing enzymes utilized in fatty acid biosynthesis. For each subunit condensation some of these can be 'edited out' such that intermediate reduction states are achieved. Therefore ketones, alcohols and double bonds as well as fully reduced methylene carbons are all characteristic features of the higher polyketides.

In this chapter the important Actinomycete antibiotic classes are discussed with particular reference to their biosynthetic origin. The diversity of structures which exist illustrates the ingenuity that the Actinomycetes display in their ability to evolve novel polyketide synthases. This chapter also highlights the extensive structural and stereochemical correlations which are apparent between various antibiotics and which are clearly indicative of common sequences of enzymatic activity and therefore substantial genetic homology between the producing strains. In this regard, information is beginning to emerge from the genetic analysis of producing strains which is providing clues to the nature, organization and control of the synthase enzymes (see section 4.8.2).

4.1 TETRACYCLINE ANTIBIOTICS

A variety of modified tetracyclic compounds have been characterized from *Streptomyces* strains, and their biosynthetic origins can be rationalized and unified by the classic folding of polyketide precursors. Pre-eminent among these compounds are the tetracyclines, which are used extensively as broad spectrum antibiotics. The parent compound of this class is tetracycline (4.6) a product of several *Streptomyces* spp. Other clinically significant analogues are oxytetracycline (4.7) and chlorotetracycline (aureomycin) (4.10) [1]. Intermediates on the biosynthetic pathway to the tetracyclines have been isolated and identified from blocked mutants of the producing strains [2,3]. The earliest identifiable intermediate is 6-methylpretetramide (4.4), which is itself transformed into the parent antibiotics in co-synthesis studies with the wild type strain. 6-Methylpretetramide (4.4) is probably the product of the polyketide synthase operating during tetracycline biosynthesis. A clue to the nature of an intermediate bound to the polyketide synthase comes from the isolation of methylanthrone (4.2), a tricyclic metabolite secreted by a mutant of *Streptomyces aureofaciens* [4]. This compound is not bio-transformed into the parent antibiotics but appears to be a shunt metabolite, possibly derived after hydrolysis and non-enzymatic oxidation of an enzyme-bound species such as (4.1).

The 6-nortetracyclines, such as (4.9), are devoid of a C-6 methyl group and are products of mutants lacking the appropriate methyltransferase activity. The absence of this functionality does not appear to preclude normal processing of this modified structural skeleton by the remaining biosynthetic enzymes, and an analogous pretetramide intermediate (4.5) and oxidized shunt product (4.3) of the nor-series have been identified in selected mutants [5].

The tetracycline antibiotics employ an intact C_3 malonyl group contributing C-1, C-2 and the carboxyamide carbon [6]. This starter group is rare, but not unique to the tetracylines, and has been identified in *Streptomyces* antibiotic cycloheximide (4.11) [7]. Common to both cases is the presence of an amide nitrogen, and there has been speculation that the true starter unit is actually malonamyl-CoA. The incorporation of malonamyl-CoA into chlorotetracycline has, however, proved inconclusive and the true identity of this starter unit remains ambiguous [8]. The occurrence of the 2-acetyl-2-decarboxamido (4.8) analogue of oxytetracycline (4.7) from *Streptomyces rimosus* can be rationalized in terms of an acetoacetyl-CoA replacing the C_3 starter unit.

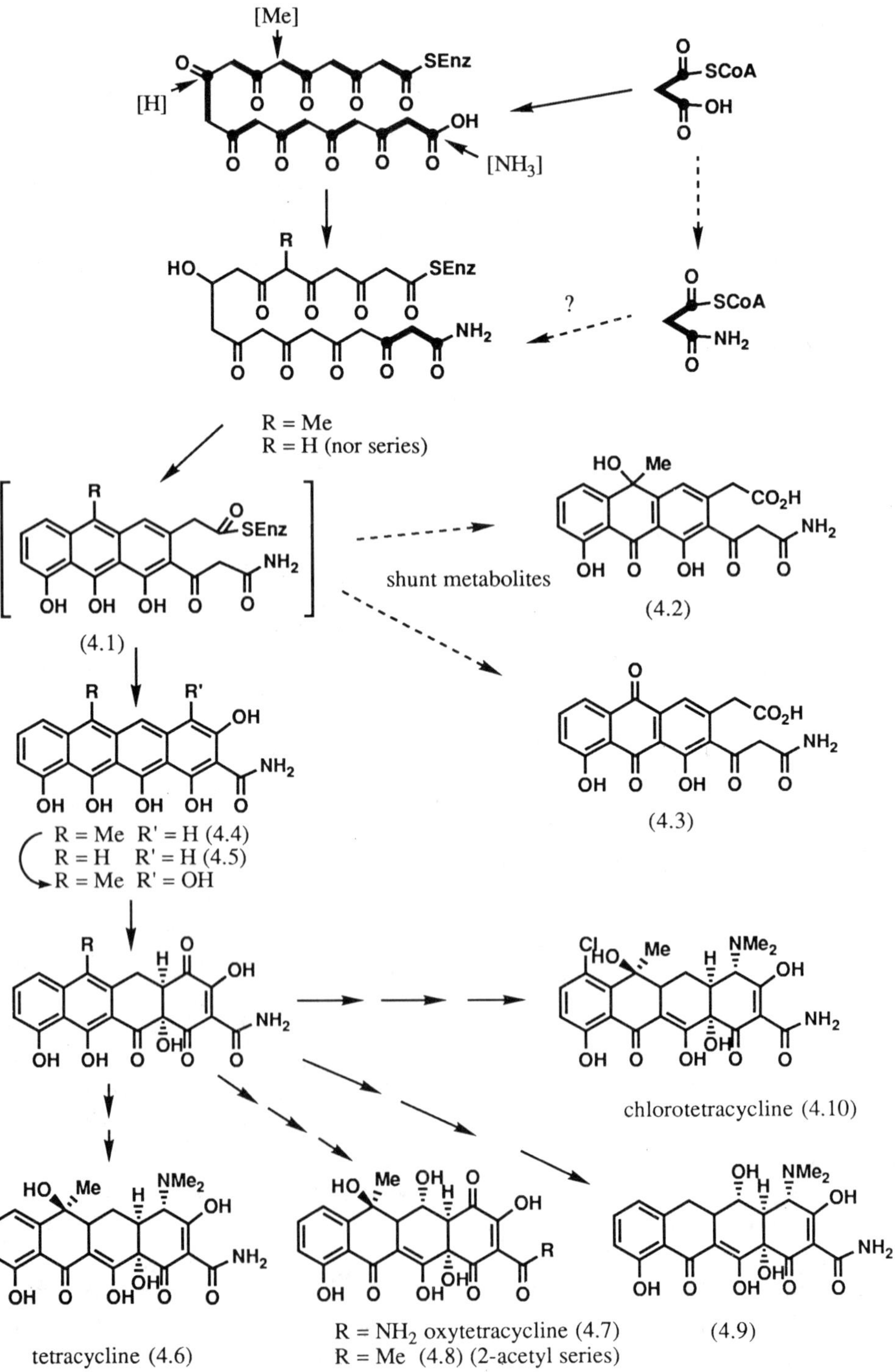
[Me]
[H]
[NH3]
SEnz
SCoA
OH
NH2
?
R = Me
R = H (nor series)
shunt metabolites
(4.1)
(4.2)
(4.3)
R = Me R' = H (4.4)
R = H R' = H (4.5)
R = Me R' = OH
chlorotetracycline (4.10)
tetracycline (4.6)
R = NH2 oxytetracycline (4.7)
R = Me (4.8) (2-acetyl series)
(4.9)

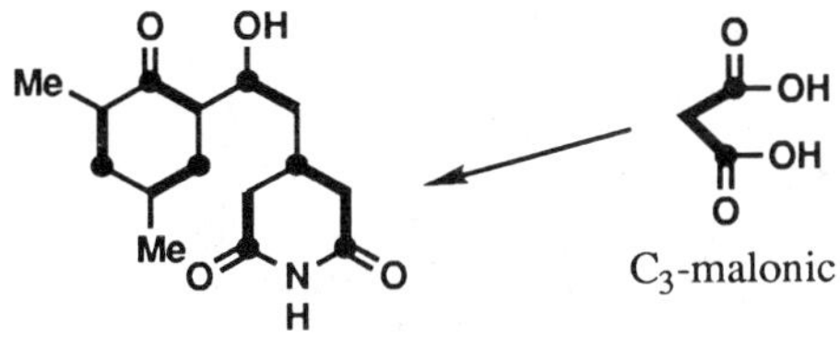

cycloheximide (4.11)

C_3-malonic acid

It is interesting to draw a comparison between the tetracyclines and the fungal metabolite viridcatumtoxin (4.12) from *Penicillium expansum.* [9]. Both systems have a similar polyketide derived skeleton and contain a carboxyamide functionality indicative of a similar starter unit. Alkylation occurs at the same site in each case, the C_{10} moiety of viridcatumtoxin originating from farnesyl pyrophosphate. On closer analysis, however, after incorporation of [1,2-$^{13}C_2$]-acetate, it has emerged that the polyketide folding pattern is not synchronous with the tetracyclines and [1,2,3-$^{13}C_3$]-malonic acid is not incorporated as an intact unit into (4.12). The carboxamide functionality is at the terminus rather than at the start of this chain. It has been suggested that the C-3 carbon of viridicatumtoxin (4.12) may originate from C-2 of glyoxylate, C-1 being lost after decarboxylation, as it is not labelled by acetate or bicarbonate. Incorporation of oxygen-18 from [1-^{13}C, $^{18}O_2$]-acetate into the 4α hydroxyl of viridicatumtoxin precludes the intermediacy of an aromatic tetracene intermediate, analogous to 6-methylpretetramide (4.4). The apparent similarities between this fungal metabolite and the bacterial tetracyclines illustrates that generalizations on the details of polyketide assembly cannot be made with confidence when comparisons are drawn between different taxonomic groups.

4.2 ANTHRACYCLINONE ANTIBIOTICS

Daunomycin (4.13) (*Streptomyces peucetius*), the antitumor antibiotic adriamycin (4.14) (produced by a mutant of *S. peucetius*), carminomycin I (4.15) (*Actinomadura carminata*) and rutilantinone (ε-pyrromycinone) (4.16) are representatives of a large class of anthracyclinone antibiotics which originate from biogenetically related polyketide synthases employing a propionate (C_3) starter unit [10,11]. For this series, an enzyme-bound tricyclic intermediate such as (4.17) can be proposed which is analogous to (4.1) in the tetracycline pathway. The aldol condensation required to generate the final ring system occurs in the opposite sense, however, i.e. from C-2 to the carbonyl of the starter subunit rather from C-18 to the carbonyl of the terminal subunit. Two basic aglycone skeletons exist resulting from either

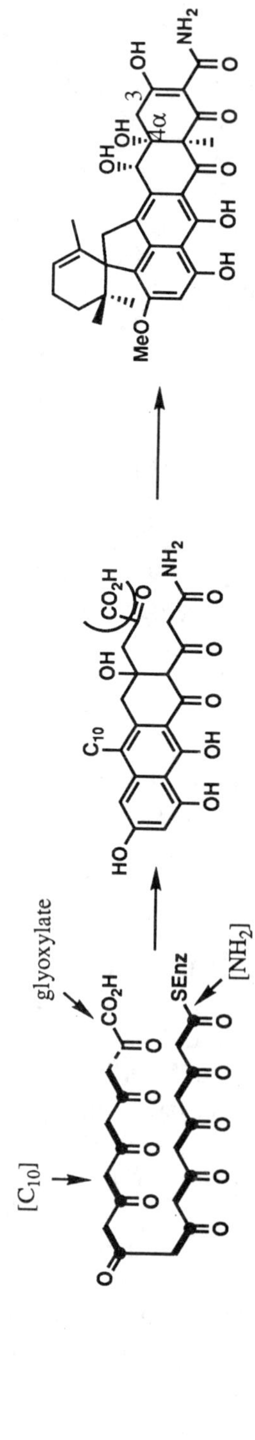
[C_{10}]
glyoxylate
CO_2H
SEnz
[NH_2]
HO
OH
OH
C_{10}
OH
CO_2H
NH_2
MeO
OH
OH
OH
OH
OH
3
4α
NH_2
OH

(4.12)

decarboxylation or retention of the terminal carboxylate group. Subsequent modifications, such as methylations, hydroxylations or glycosylations, then generate the various antibiotics.

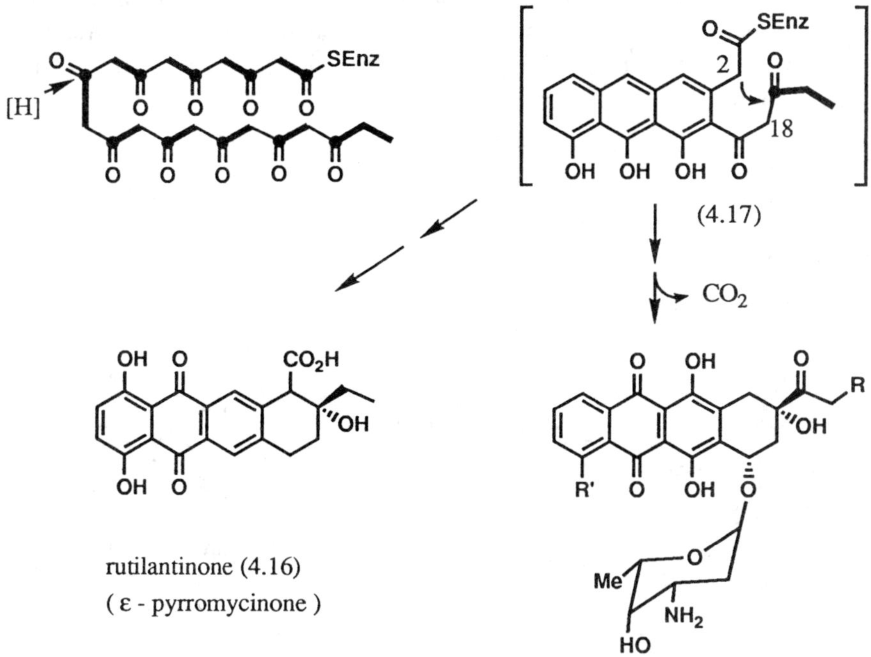

(4.17)

rutilantinone (4.16)
(ε - pyrromycinone)

R = H, R' = Me daunomycin (4.13)
R = OH, R' = Me adriamycin (4.14)
R = R' = H carminomycin I (4.15)

4.3 ISOCHROMANEQUINONE ANTIBIOTICS

The isochromanequinones are a class of pigmented *Streptomyces* antibiotics typified by medermycin (4.18), granaticin (4.19) and the dimeric structure actinorhodin (4.20). They are acetogenins derived in the classic manner from the folding of an acetate/malonate polyketide chain [12]. The producing strains of these antibiotics have attracted attention recently as the source of the first 'hybrid' antibiotics. Hybrids are generated through genetic engineering by introducing some of the genes responsible for the production of one antibiotic into a host organism with a complete set of genes coding for the synthesis of a second antibiotic. For example, the entire set of genes from *Streptomyces coelicolor* responsible for the production of actinorhodin (4.20) was cloned, and then fragments of the cloned DNA were introduced into the medermycin (4.18) producer *Streptomyces* sp. [AM-7161]. Mederrhodin-A (4.21), which possesses all the structural features of medermycin (4.18) as well as an extra hydroxyl group at C-8 characteristic of actinorhodin

(4.20), was secreted from one of the engineered organisms [13,14]. Similarly, dihydrogranatirhodin (4.22) was generated as a hybrid of actinorhodin (4.20) and granaticin (4.19). The hybrid antibiotic has structural and stereochemical features associated with both parent antibiotics. This is exciting technology which holds promise in the future for generating new antibiotics and even customizing existing antibiotics to include desirable structural features and modifications.

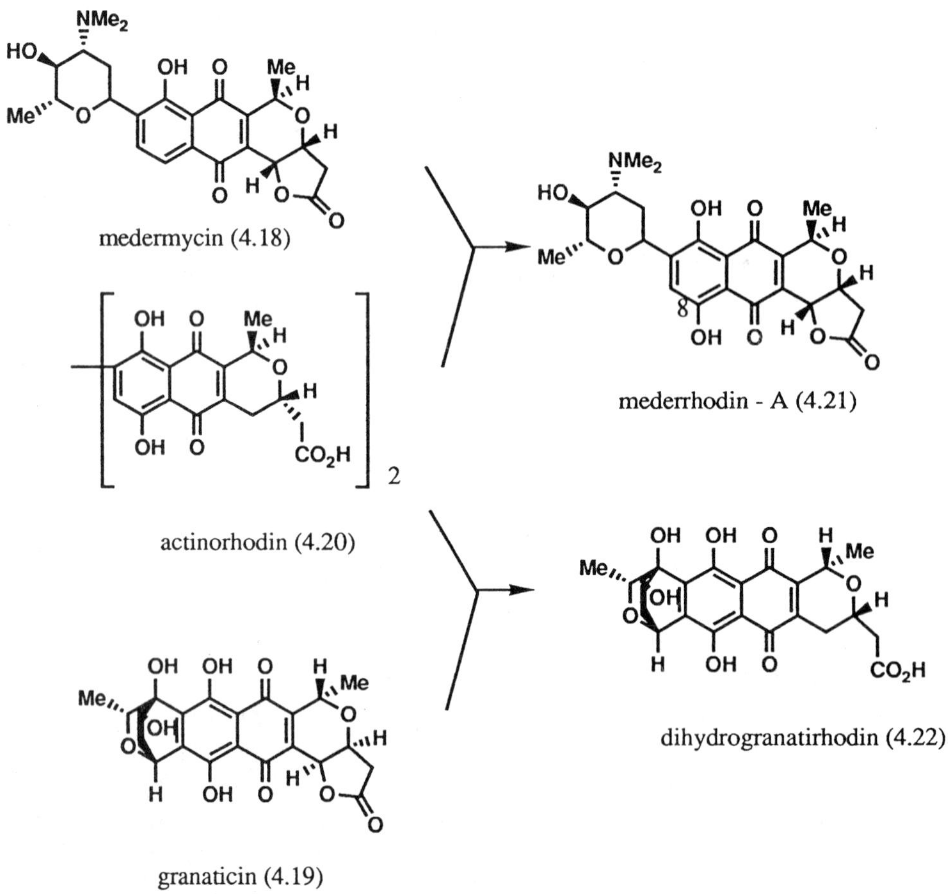

4.4 HIGHER ORDER POLYKETIDES

4.4.1 Subunits

Up until this point we have largely considered 'acetogenic' polyketides, i.e. those that utilize malonate for chain extension, although we have seen a variety of alternative starter groups employed in fatty acid (see section 1.3.1) and fungal metabolite (see section 3.3.2) biosynthesis. Actinomycetes do, however, have an amazing capacity to biosynthesize polyketides which incorporate propionate, and to some extent butyrate, into their structural skeletons. These higher polyketides are typified by the macrolide (e.g. erythromycin-A (4.23)), polyether (e.g. monensin-A (2.24)) and ansamycin (e.g. rifamycin-B (4.25)) classes of antibiotics. The regular array of methyl groups on the erythromycin-A (4.23) aglycone led Gerzon [15] and Woodward [16] to propose a 'propionate rule', analogous to Birch's 'acetate hypothesis', which suggested that the erythromycin skeleton was derived from a propionyl-CoA starter unit and six methylmalonyl-CoA subunits. Subsequent biosynthetic investigations have borne this out [17]. The erythromycins are unusual Actinomycete metabolites in being derived entirely from propionate (C_3) subunits, although there are an increasing number of polypropionate metabolites being identified in marine organisms (see, for example, section 5.4). More typically, however, for the Actinomycetes, the higher polyketides arise from combinations of C_2 (acctatc), C_3 (propionate) and C_4 (butyrate) subunits. The aglycone of the 16-membered macrolide antibiotic, tylosin (4.26) is representative and is constructed from two acetate

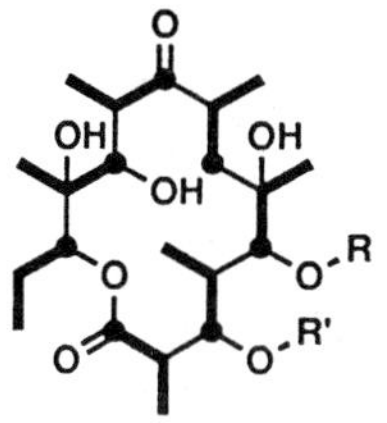

$P_1P_2P_3P_4P_5P_6P_7$

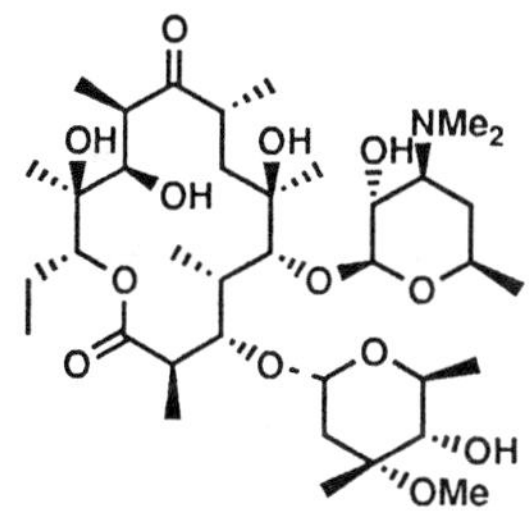

erythromycin-A (4.23)

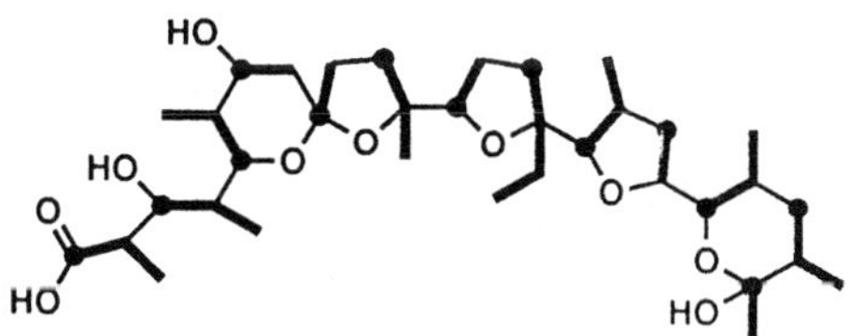

$P_1P_2P_3A_4A_5P_6A_7B_8P_9A_{10}P_{11}P_{12}A_{13}$

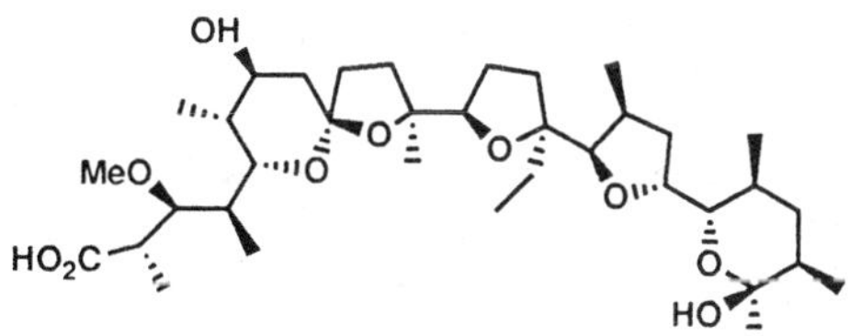

monensin-A (4.24)

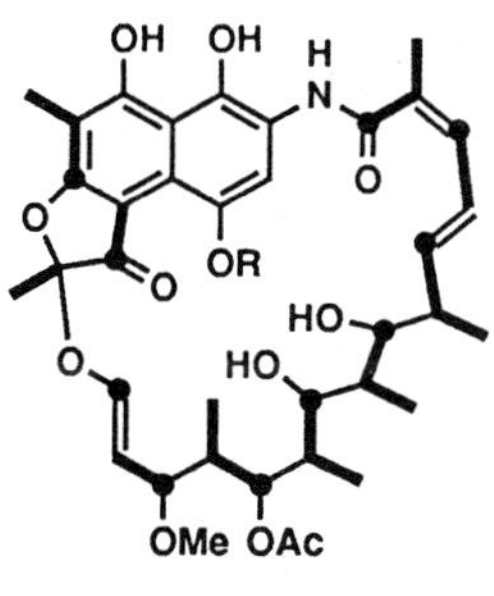

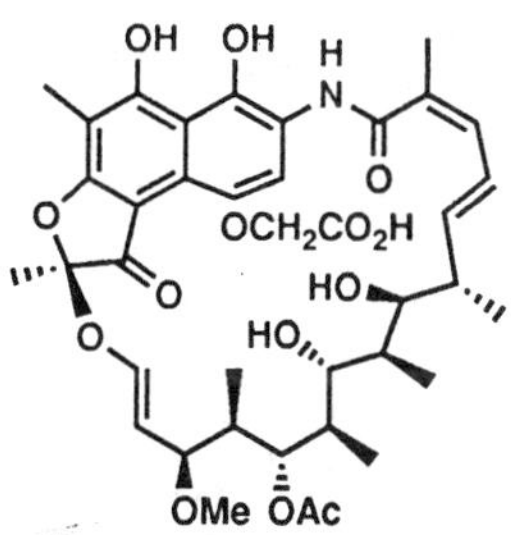

rifamycin B (4.25)

C_7N unit-$P_2A_3P_4A_5P_6P_7P_8P_9A_{10}P_{11}$

subunits, five propionate subunits and one butyrate subunit. Similarly the polyether antibiotic monensin-A (4.24) is assembled from five acetate, seven propionate and one butyrate subunits. Interestingly, a study [18] on the 16-membered macrolide leucomycin-A_3 (4.27), from *Streptoverticillium kitasatoensis*, has revealed that C-3 and C-4 of this antibiotic do not derive from acetate. They may, however, have an origin in glycolate.

Nonactin (4.28) (*Streptomyces griseus*) is the parent compound of a series of analogous ionophore antibiotics which exist as cyclic tetramers of alternating enantiometric forms of nonactic acid and its homologues. The involvement of intact succinate as a contributory C_4 subunit to the backbone of nonactin is unusual and interesting [19,20]. The framework is derived from one succinate, one propionate and two acetate subunits. Assembly of nonactin is non-classical, as chain extension proceeds in two directions from succinate. A decarboxylation occurs prior to the cyclization to generate the tetrahydrofuran ring system, isolating C-7 as a single carbon derived from C-2 of acetate.

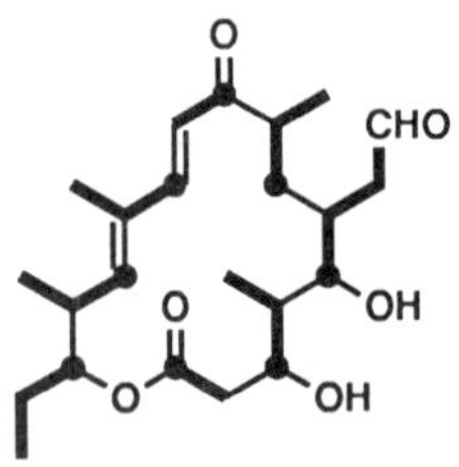

$P_1P_2P_3A_4P_5B_6P_7A_8$

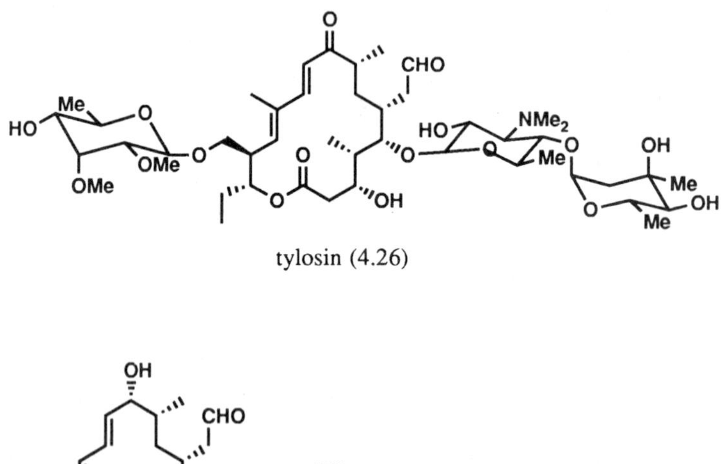

tylosin (4.26)

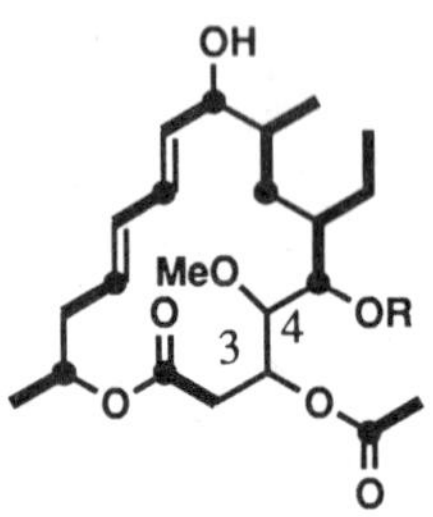

$A_1A_2A_3A_4P_5B_6G_7A_8$

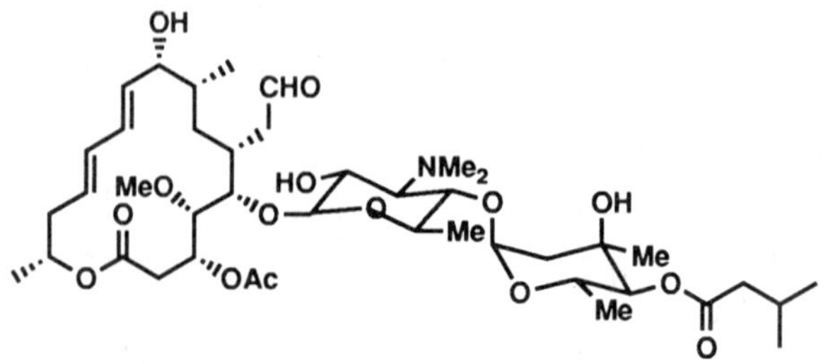

leucomycin - A_3 (4.27)

The utilization of pyruvate as a starter group in polyketide assembly has been observed in the biosynthesis of coronatine (4.29) from *Pseudomonas syringae*. This bacteria is not an Actinomycete but it is noteworthy for this unique starter group [21].

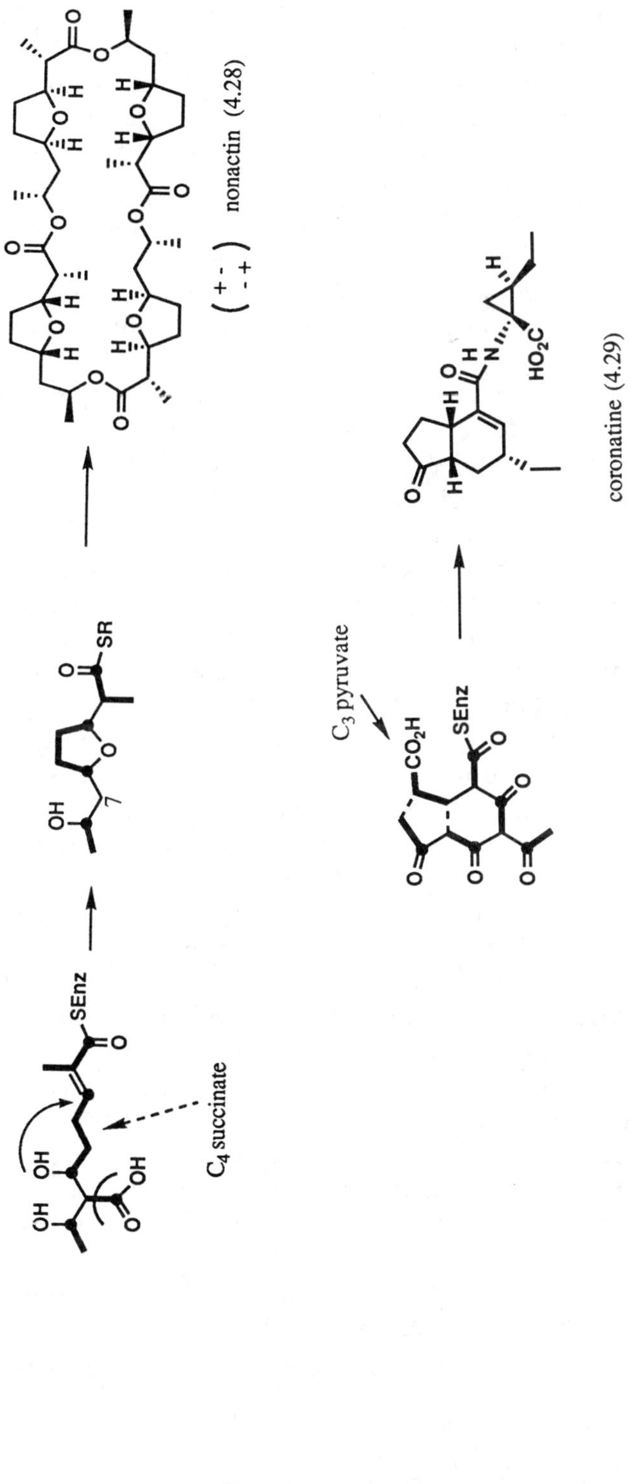

nonactin (4.28)

coronatine (4.29)

4.4.2 Polyketide chain assembly

The condensation of acetate, propionate and butyrate subunits in the higher polyketides is believed to have analogy with classical fatty acid biosynthesis. Thus, like acetyl-CoA in the fatty acid systems, propionyl-CoA and butyryl-CoA are activated by carboxylase enzymes to generate methylmalonyl-CoA and ethylmalonyl-CoA respectively. The polyketide synthases are programmed to control the selection of each substrate such that a decarboxylative condensation of the appropriate precursor can take place with the developing polyketide chain. A consequence of the incorporation of methyl- and ethyl-malonyl-CoAs is the introduction of methyl and ethyl functionality respectively onto the polyketide backbone. Furthermore these alkyl groups adopt either a D- or L-configuration with respect to the direction of chain assembly.

Propionyl-CoA carboxylase from the erythromycin-A producer *Saccharopolyspora erythraea* (formerly termed *Streptomyces erythreus*) has been purified to homogeneity and shown to mediate the conversion of propionyl-CoA to (2*S*)-methylmalonyl-CoA (4.30) [22]. (2*S*)-Methylmalonyl-CoA (4.30) is then processed by the polyketide synthase after a transesterification onto an acyl carrier protein (ACP). The evidence suggests that the next step, the decarboxylative condensation, proceeds with *inversion of configuration* consistent with fatty acid biosynthesis. This information has been obtained from biosynthetic investigations on several metabolites employing propionate, carrying deuterium at C-2 [23–25]. In all of the cases studied, deuterium was retained only at those methine sites adjacent to D-methyl groups and was absent at methine sites adjacent to L-methyl groups in the isolated antibiotics. These observations can be rationalized as follows. The decarboxylative condensation of (2*S*)-methylmalonyl-ACP occurs with inversion of configuration to give a D-α-methyl-β-keto-thioester (4.31). This would lead to retention of deuterium at the D-methyl centres. The L-methyl centres presumably arise after epimerization of a D-centre and the deuterium would therefore be lost to the medium. This epimerization can occur at one of two stages during the process. Possibly (2*S*)-methylmalonyl-CoA (4.30) is epimerized to (2*R*)-methylmalonyl-CoA (4.32) and the polyketide synthase then accepts each diastereoisomer to order. In order to probe this possibility attempts have been made to generate (2*R*)-methylmalonyl-CoA *in vivo* (see section 4.4.3), carrying deuterium at C-2 [25,26]. These have failed to result in the incorporation of the deuterium isotope at the L-methine centres, although this failure can be attributed to extensive washout of the deuterium label rather than complete rejection of (2*R*)-methylmalonyl-CoA. Alternatively, (2*S*)-methylmalonyl-CoA (4.30) is the only diastereoisomer utilized by the synthase for the condensation and the resultant D-α-methyl-β-keto-thioester (4.31) is epimerized to generate an L-α-methyl-β-keto-thioester

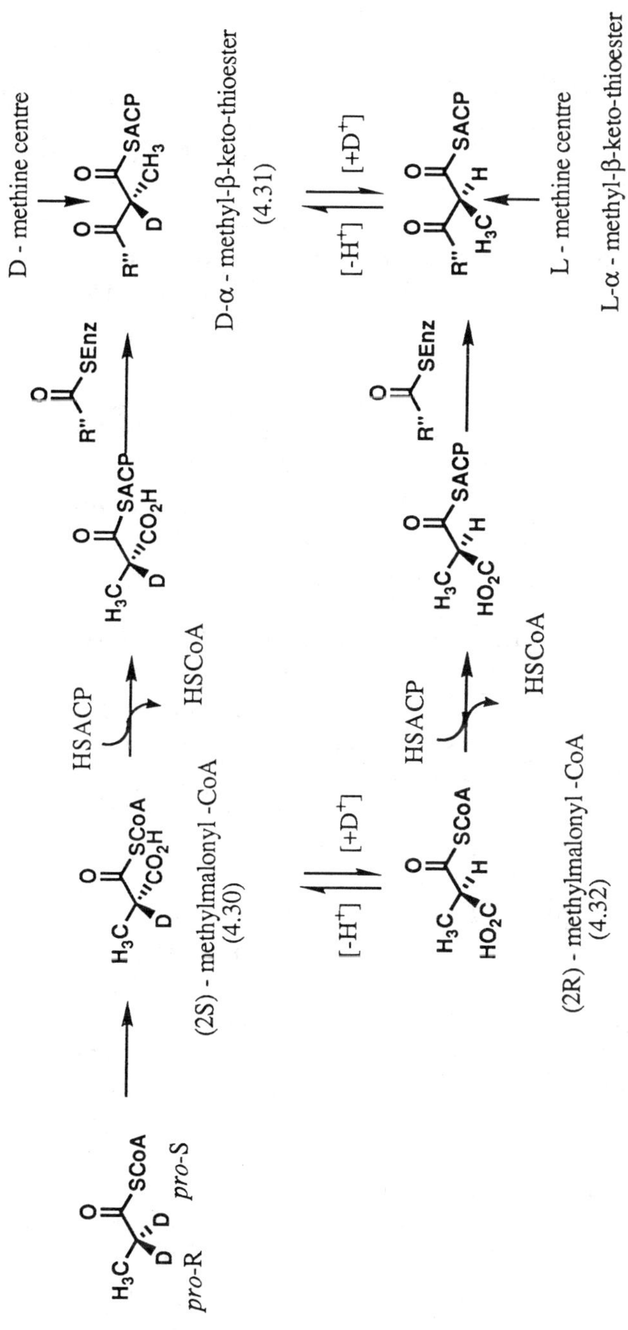
pro-R
pro-S
(2S) - methylmalonyl -CoA
(4.30)
HSACP
HSCoA
D - methine centre
D-α - methyl-β-keto-thioester
(4.31)
[-H+]
[+D+]
(2R) - methylmalonyl -CoA
(4.32)
L - methine centre
L-α - methyl-β-keto-thioester
(4.33)

(4.33). The delineation of these two processes, i.e. a methylmalonyl-CoA epimerase or an epimerase acting periodically on the polyketide chain, remains an interesting and unresolved problem in polyketide biosynthesis. The stereochemistry of butyrate metabolism has been studied in one instance [26]. The resultant incorporation of deuterium from butyrate carrying deuterium at C-2 demonstrated that the stereochemical processing of butyrate and ethylmalonyl-CoA is the same as described above for propionate and methylmalonyl-CoA.

4.4.3 The sources of propionate and butyrate presursors

Those bacterial polyketide metabolites which incorporate propionate require a metabolic source of methylmalonyl-CoA. Methylmalonyl-CoA can be generated directly from propionyl-CoA by propionyl-CoA carboxylase, and indeed this activity has been demonstrated in the erythromycin producer *S. erythraea*, giving rise specifically to (2*S*)-methylmalonyl-CoA (4.30) [22] (see section 4.4.2). The (2*R*)-diastereoisomer (4.32) can be accessed directly from succinate by the action of the coenzyme-B_{12}-dependent enzyme methylmalonyl-CoA mutase, and incorporation studies involving [^{13}C]-succinates appear to support the presence of this activity in several antibiotic-producing strains [25,26,27].

Butyrate can clearly be derived from the β-oxidation of fatty acids, but it also has an amino acid source in Streptomycetes. It had been recognized for a number of years that butyrate is efficiently converted into the propionate subunits of some polyketide antibiotics [28], although it is only recently that the details of the pathway linking propionate and butyrate in Streptomycetes have been revealed [29]. The key intermediate is isobutyryl-CoA (4.36). Isobutyrate (4.35), a catabolic product of L-valine (4.34), is efficiently incorporated into both the propionate and butyrate subunits of the macrolide and polyether antibiotics [30,31]. The incorporation of isobutyrate (4.35) into the propionate subunits can be rationalized by conversion to its coenzyme-A ester and then oxidation of one of the methyl groups of isobutyryl-CoA (4.36). Studies probing the stereochemistry of this process have revealed that it is the *pro-S* methyl group of isobutyrate that is oxidized and lost as carbon dioxide, and therefore this pathway gives rise directly to (2*S*)-methylmalonyl-CoA (4.30) prior to incorporation into the propionate subunits [29].

The incorporation of isobutyrate (4.35) into the butyrate subunits of the polyketide antibiotics involves the action of a recently characterized enzyme, isobutyryl-CoA mutase [32]. This enzyme utilizes coenzyme-B_{12} as a cofactor and mediates the reversible rearrangement of isobutyryl-CoA (4.36) and butyryl-CoA (4.37). The stereochemistry course of this rearrangement has been evaluated in *Streptomyces cinnamonensis*, and the C-1 carboxyl carbon

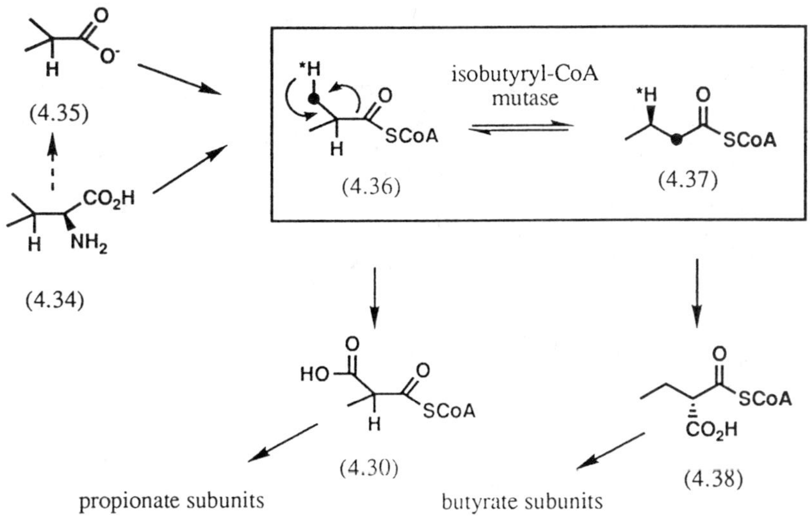

of isobutyryl-CoA migrates to the *pro-S* methyl group of isobutyryl-CoA [28]. This methyl group then becomes C-2 of butyryl-CoA. Consistent with other coenzyme-B_{12} enzymes, a vicinal interchange process operates during the isomerization, i.e. a proton from the *pro-S* methyl group of isobutyryl-CoA is relocated at C-3 of butyryl-CoA, predominantly in the 3-*pro-R* position. Therefore the carboxyl carbon is interchanged with the proton of the methyl group with overall *retention* of configuration. The butyryl-CoA (4.37), once formed, is probably activated then by carboxylation to (2*S*)-ethylmalonyl-CoA (4.38) for incorporation into the butyrate subunits of the antibiotics, although as yet a specific butyryl-CoA carboxylase has not been identified from these systems.

Isobutyryl-CoA mutase activity has only been detected so far in Streptomycetes. The enzyme may be important for linking the catabolic pathways of the amino acids and fatty acids and for controlling the levels of the C_3 and C_4 pools, such that the organisms can channel appropriate precursors efficiently into antibiotic biosynthesis.

4.4.4 Macrolide antibiotics — biosynthesis

The biosynthesis of the 14- and 16-membered macrolide antibiotics erythromycin A and tylosin have received a great deal of attention in view of their commercial significance. In each case blocked mutants of the producing strains have been isolated and cultured which secrete the aglycones (sugar-free nuclei) of these antibiotics. Thus 6-deoxyerythronolide-B (4.39) [33] and tylactone (4.40) would appear to be the products of the polyketide synthases. Similarly, platenolide (4.41) has been isolated from a blocked mutant of *Streptomyces platensis* [34]. The subsequent elaboration of this

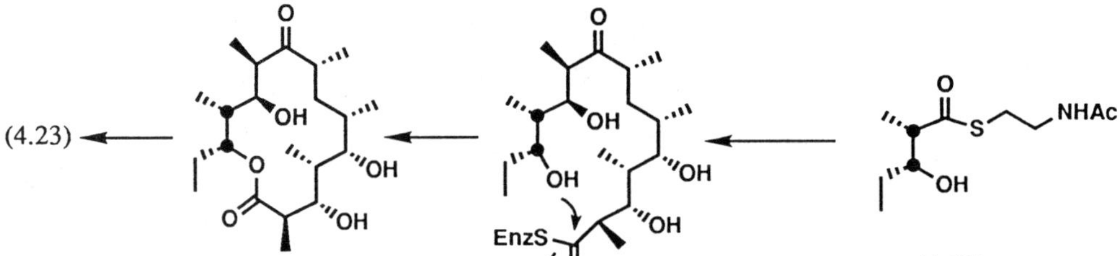

6-deoxyerythronolide (4.39) (4.49) (4.48)

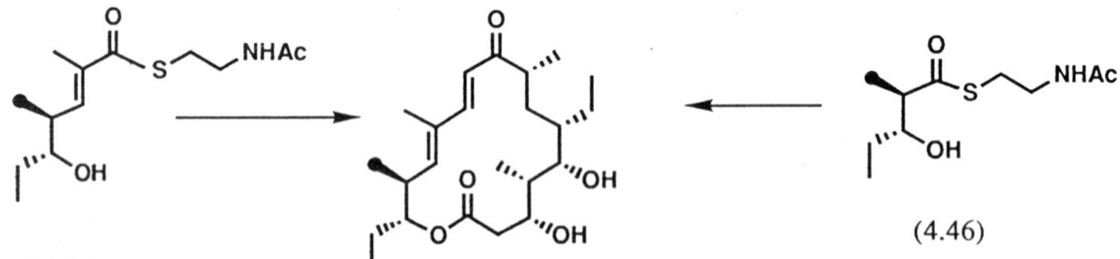

(4.47) tylactone (4.40) (4.46)

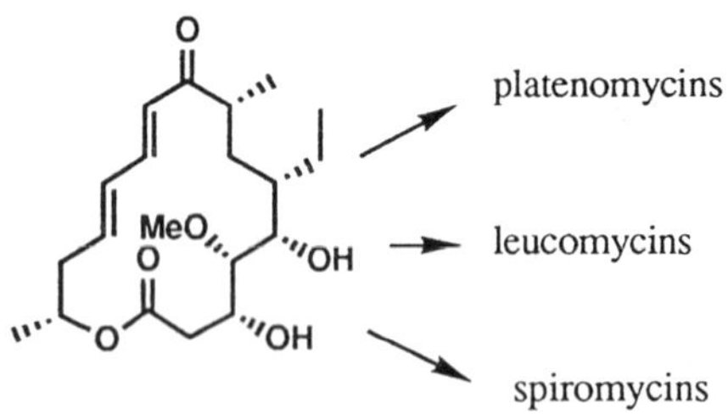

platenolide (4.41)

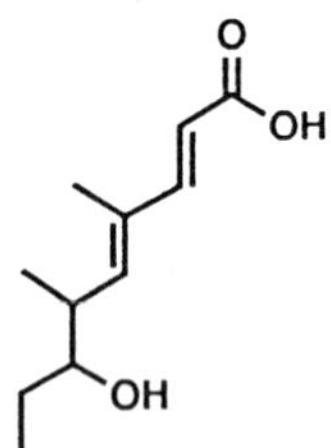

(4.45)

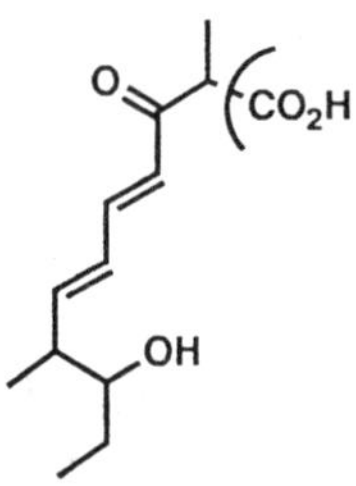

(4.42)

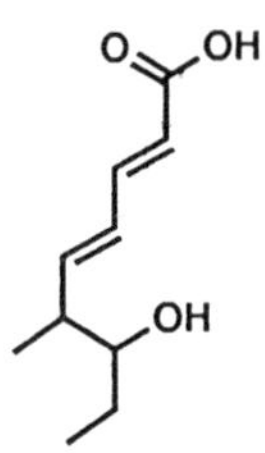

(4.43)

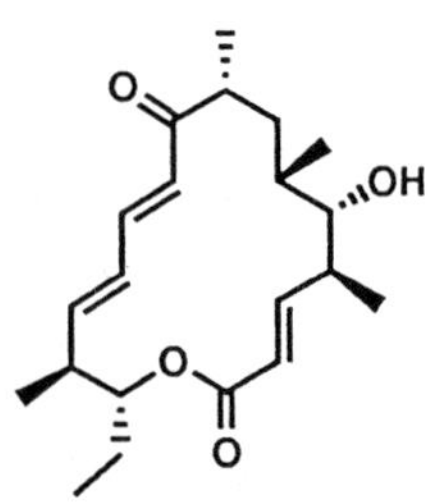

mycinolide IV (4.44)

aglycone gives rise to a large number of antibiotics of the platenomycin, leucomycin and spiromycin groups, and their origins can clearly be interpreted in terms of a common polyketide synthase which secretes platenolide, operating in the various producing strains.

Until recently the identification of intermediates between the C_2–C_4 carboxylate precursors and the assembled algycones of the macrolides remained elusive. However, fragments such as (4.42) and (4.43) of mycinolide IV (4.44), the aglycone of mycinamicin macrolides, have now been identified from blocked mutants of *Micromonospora griseorubida*. Similarly fragments such as (4.45) have been identified from mutants of *Streptomyces fradiae* which secretes tylactone (4.40) [36] and it would appear that these partially elaborated segments are off-loaded at an intermediate stage from the mycinamicin and tylactone synthases in the mutant strains [35]. Their identification has mechanistic implications which are discussed below.

Some of the most interesting and revealing biosynthetic experiments on the macrolide antibiotics have involved the intact incorporation of partially elaborated aglycone fragments into both the tylosin aglycone tylactone (4.40) [37] and erythromycin-A (4.23) [38]. The *S*-acyl moieties of the ^{13}C-labelled compounds (4.46) and (4.47) were incorporated into tylactone (4.40) when incubated with the tylactone-producing mutant of *Streptomyces fradiae*. Similarly, the doubly labelled $^{13}C_2$ compound (4.48) was incorporated intact into erythromycin-A (4.23). See also nargenicin A in section 4.6. The success of these experiments relied on the preparation of *N*-acylcysteamine thioesters rather than presenting the free carboxylic acids or carboxylic acid esters to the producing organisms. Unlike their carboxylate counterparts these thioesters appear to exchange with (or replace?) the ACP of the polyketide synthase and intercept the polyketide assembly process. These experiments coupled with the isolation of the aglycone fragments (4.42), (4.43) and (4.45) of mycinamicin and tylactone forcefully demonstrate that the macrolide aglycones are constructed by a *processive* mechanism. This term describes a process where the final reduction level of the β-ketone functionality is adjusted and the stereochemical configuration set immediately after each condensation event and prior to the next condensation. Genetic analysis of the erythromycin system adds further support to this mechanism (see section 4.8.3).

Evidence from oxygen-18 studies support a macrolactonization as the last step in aglycone biosynthesis. For the erythromycins and tylosin it has been demonstrated that all of the oxygens of the aglycones are derived intact from their carboxylate precursors [39,40]. Significantly, the lactone carbonyl oxygen and the bridging oxygen of the lactone originate from the carboxylate oxygens of *distinct* subunits, consistent with the macrolactonization of a fully elaborated thioester such as (4.49) in erythromycin-A biosynthesis. Subsequent modifications leading to the parent antibiotics involve hydroxylations

(O from molecular oxygen), introduction of pendent glycosyl groups (from D-glucose) and methylations (CH_3 from L-methionine).

4.4.5 Macrolide stereochemistry

A striking feature of the macrolide antibiotics is the stereochemical homology which exists between the antibiotics in this group. For example, when the 12-membered macrolide methymycin (4.50) and the 14-membered macrolide erythromycin-A (4.23) are represented linearly by the Fischer convention and compared, the absolute stereochemistries of the oxygen and alkyl functionality along their respective backbones match exactly. In 1965 Celmer defined a configurational model (4.51) which summarizes the absolute stereochemistry of *all* the 12- and 14-membered macrolides [41]. Insertion of a two-carbon unit between C-10 and C-11 extends the model to include the 16-membered macrolides as illustrated for mycinamicin-IV (4.52). Although the correlations remain substantial, one asymmetric centre becomes inconsistent. When alkyl functionality is present at C-14 of a 16-membered macrolide, the stereochemistry is found to be opposite to that predicted by Celmer's model. This is, however, a regular inconsistency and the absolute stereochemistry of all the 16-membered macrolide antibiotics is uniform [42]. A further stereochemical homology of the macrolides is evident in the glycoside residues. All glycosidic linkages at C-3 possess an (α-L) configuration and those at C-5 possess a (β-D) configuration. The high level of stereochemical homology between these structurally complex metabolites indicates that the individual polyketide synthases have very similar characteristics and probably possess a very close biogenetic relationship.

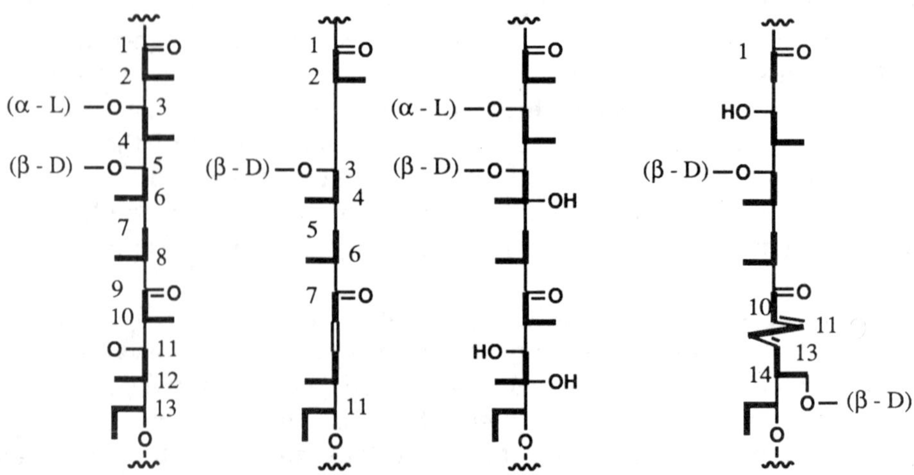

Celmer model (4.51) | methymycin (4.50) | erythromycin A (4.23) | mycinamicin IV (4.52)

4.4.6 The avermectins and milbemycins

The avermectins and milbemycins are 16-membered macrolides of *Streptomyces avermitilis* and *Streptomyces hygroscopicus* respectively. They display no significant antibacterial activity but are potent antiparasitic agents and appear to interfere with invertebrate neurotransmission. These macrocyclic lactones are higher polyketides derived largely from acetate and propionate subunits. Many analogues of each series have been identified since they were first isolated in the early 1980s [43,44]. These analogues arise in part from the utilization of various starter units by the producing synthases. For example milbemycin-α_1 (4.53) incorporates an acetate starter unit, whereas milbemycin-D (4.54) is primed from isobutyrate, itself originating from L-valine catabolism [45]. Similarly the *sec*-butyl group of avermectin-$A_{1\alpha}$ (4.55) is derived from the incorporation of 2-methylbutyrate, a product of L-leucine metabolism [46]. All of the oxygen atoms of these aglycones including that at C-13 of the avermectins appear to derive from their carboxylate precursors, except the C-6 ether oxygen which probably originates from molecular oxygen. In a similar manner to the macrolide antibiotic aglycone studies (see section 4.4.4) macrolactonization can account for ring construction. Although there are clear biosynthetic similarities between these metabolites and the macrolide antibiotics, there does not appear to be any obvious stereochemical homologies.

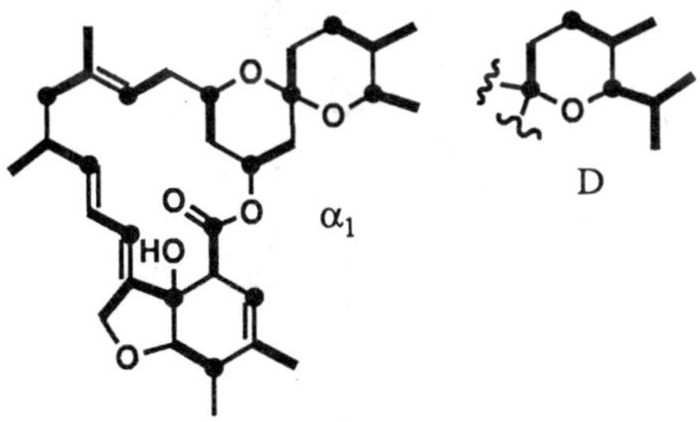

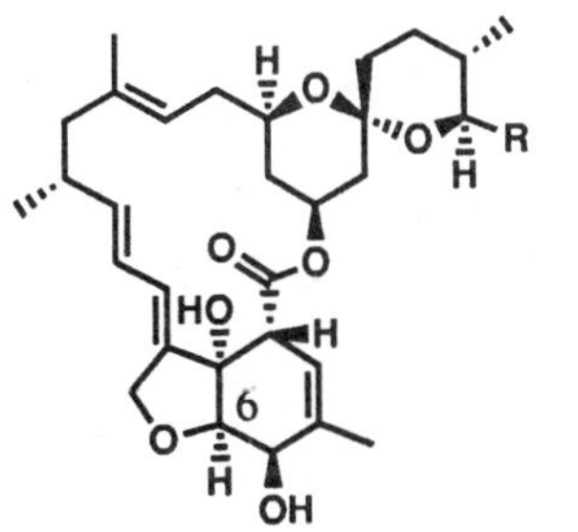

R = Me milbemycin α_1 (4.53)
R = *i*Pr milbemycin D (4.54)

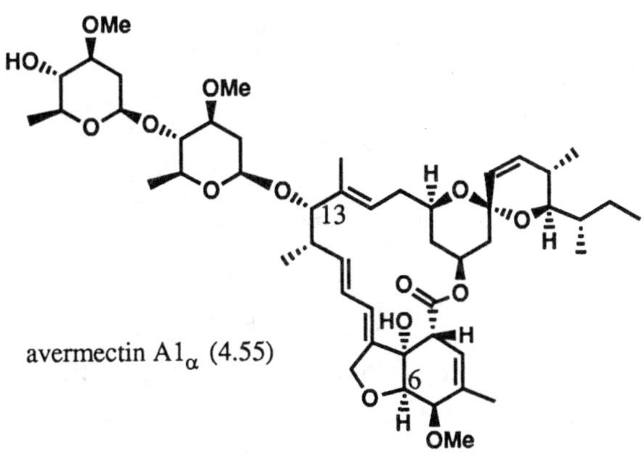

avermectin $A1_\alpha$ (4.55)

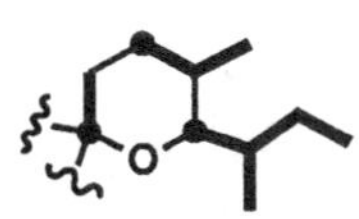

D

4.4.7 The ansamycin antibiotics

The ansamycin antibiotics are typified by the rifamycins and the streptovaricin classes. The 'ansa' bridge system of rifamycin-B (4.25) has a polyketide origin and is derived from a combination of acetate and propionate subunits. Typically the ansamycins possess a naphthalenoid or benzenoid moiety which is assembled in part from 3-amino-5-hydroxybenzoic acid (a C_7N subunit). This acid is efficiently incorporated into the ansamycins and appears to serve as a starter unit for polyketide assembly. In support of this is the identification of a compound designated P8/1-OG (4.56) from mutants of the rifamycin producer *Nocardia mediterranei* [47]. The accumulation of this fragment suggests that it has been expelled from a polyketide synthase after a C_3–C_2–C_3 polyketide condensation sequence. In the rifamycins, the contiguous nature of the polyketide framework is broken by an ether linkage between C-29 and C-12, although both of these carbons derive from the same propionate unit. This clearly suggests the operation of a single polyketide synthase mediating construction of the 'ansa' bridge system and that oxygen insertion occurs after chain assembly [48].

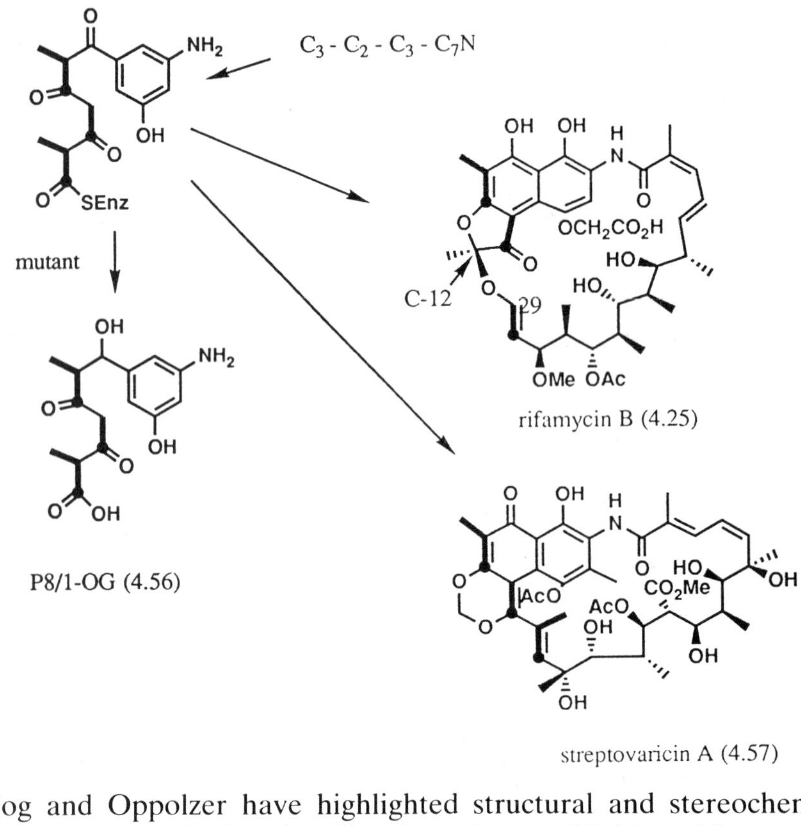

P8/1-OG (4.56)

rifamycin B (4.25)

streptovaricin A (4.57)

Prelog and Oppolzer have highlighted structural and stereochemical homologies between the functionality along the 'ansa' bridge of the ansamycins and the macrolide antibiotic aglycones [49]. These correlations are illustrated in Fig. 4.1 for rifamycin-B (4.25) and streptovaricin-A (4.57) by comparison with erythromycin-A (4.23) and the Celmer model (4.51) (see section 4.4.3). There is only one inconsistent chiral centre (asterisked) over the stereochemically defined regions of the 'ansa' bridge and the macrolide backbones. Again these homologies are indicative of considerable homology at the genetic level for the expression of the polyketide synthases (see section 4.8.3).

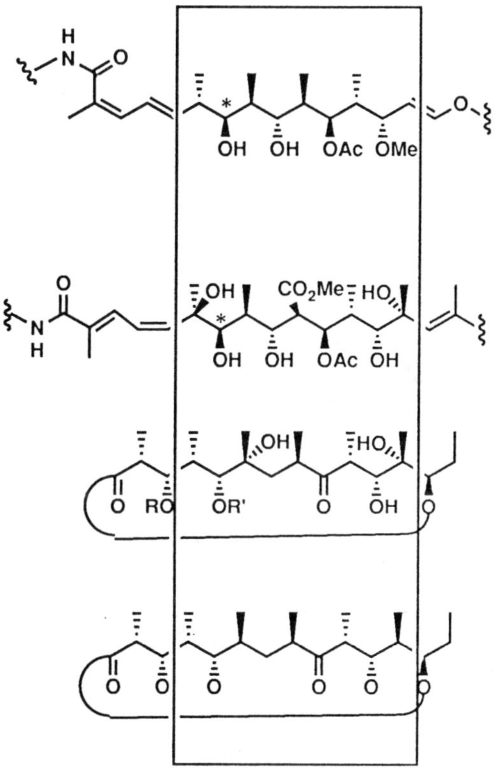

Fig. 4.1.

4.4.8 Polyene and polyhydroxy antibiotics

The polyene and polyhydroxy antibiotics are antifungal rather than antibacterial agents. They are toxic to mammals and have found limited clinical use, being used only as topical applications for fungal infections. Amphotericin-B (4.58), a product of *Streptomyces noussei*, is a typical example, containing extended conjugated and β-hydroxy systems [50]. Many analogues exist in this series with various chain lengths and extents of unsaturation and hydroxylation. The aglycones of these antibiotics originate largely from acetate, and the branching methyl groups tend to derive from the incorporation of a propionate subunit. An interesting example is fungichromin (4.59) from *Streptomyces cellulosae*, a micro-organism found in Texan soil, which is constructed from the classic assembly of a polyacetate chain incorporating one propionate subunit. At the terminus of the polyketide chain a C_8 octanoate group is specifically incorporated into fungichromin and probably originates from the β-oxidation of fatty acids [51]. The involvement of subunits longer than C_4, particularly at the terminal rather than starter site, is unusual in polyketide assembly.

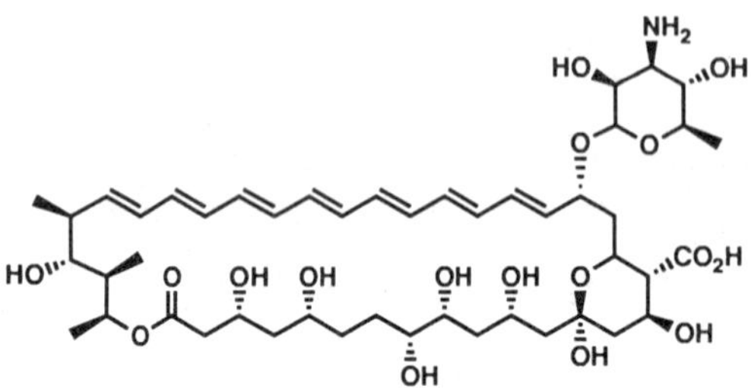

amphotericin B (4.58)

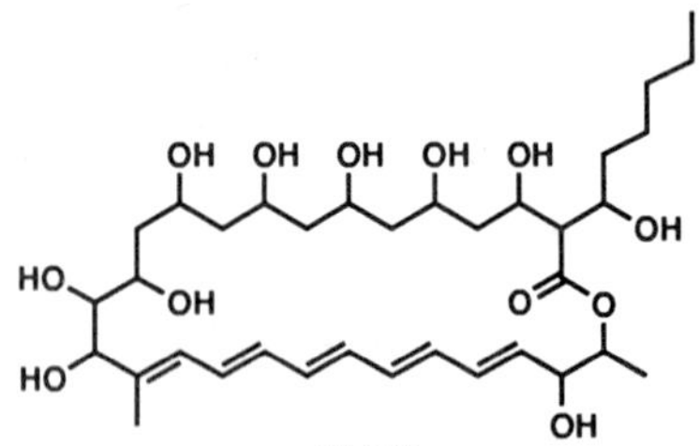

(4.59)

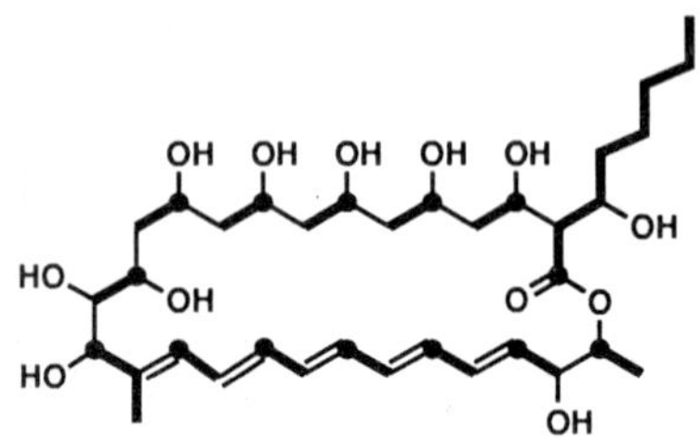

4.4.9 Polyether antibiotics — biosynthesis

The polyether antibiotics are natural ionophores. Any array of tetrahydrofuranyl and tetrahydropyranyl ring systems are arranged in these metabolites such that the oxygen atoms are suitably disposed to complex alkali metal ions in a similar manner to crown ethers. They act, therefore, by disrupting the Na^+/K^+ balance across bacterial cell membranes. Owing to their high level of toxicity they have found limited clinical use but are used as anticoccidiostats in veterinary medicine and as feedstock additives. Lasalocid-A (4.60) was among the first of this class of antibiotic to be isolated in 1951, although its structure was not fully determined until 1970 [52]. A visual analysis of the structure of lasalocid-A (4.60) suggests a polyketide origin from five acetate, four propionate and three butyrate subunits, and indeed this has been confirmed by appropriate biosynthetic experiments. The isolation of lasalocid-A and an isomeric co-metabolite, termed iso-lasalocid-A (4.61), from the fermentation broth of *Streptomyces lasaliensis* led Westley to propose a biosynthetic pathway to these metabolites involving a common intermediate diepoxide [53]. If the polyketide synthase assembled a functionalized aromatic diene such as (4.62), epoxidation of the diene could generate the diepoxide (4.63). A regiocontrolled opening of the terminal epoxide at C-23 (pathway a) or C-22 (pathway b) to terminate a ring cyclization process could account for the presence of both lasalocid-A or iso-lasalocid-A. Labelling studies employing [1-^{13}C, $^{18}O_2$]-acetate, propionate and butyrate lend support to this hypothesis, as all of the oxygen atoms of lasalocid-A are enriched with oxygen-18, except those attached to C-22 and C-23 [54]. Clearly these two oxygens derive from molecular oxygen if such a hypothesis is correct.

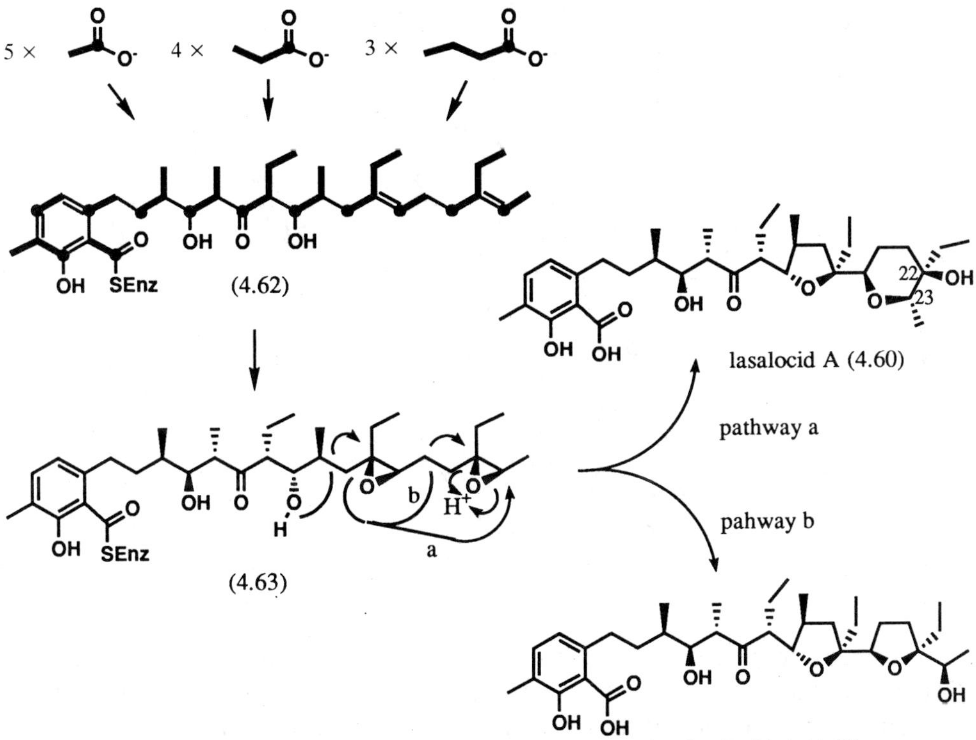

(4.62)

(4.63)

lasalocid A (4.60)

iso-lasalocid A (4.61)

Several polyether antibiotics have been biosynthesized under an atmosphere of $^{18}O_2$. Monensin-A (4.24), for example, was enriched at four sites by $^{18}O_2$ [55,56]. Three of these enrichments can be rationalized in terms of an intermediate triene (4.64) undergoing microbial epoxidation to a triepoxide (4.65). Cascade cyclization of the triepoxide would then generate the monensin-A skeleton, with the oxygen atoms from molecular oxygen ending up attached to C-13, C-17 and C-21 of monensin-A. The fourth enrichment at the C-26 hydroxyl oxygen can be accounted for by a hydroxylation occurring independently of the cyclization cascade. Complementary results have been obtained from similar $^{18}O_2$ experiments on the polyethers maduramicin [57], narasin-A [58] and lenoremycin [59]. In each case the presence of an intermediate polyepoxide can account for the observed labelling patterns, and the cumulative evidence reinforces Westley's original epoxide hypothesis for polyether biosynthesis.

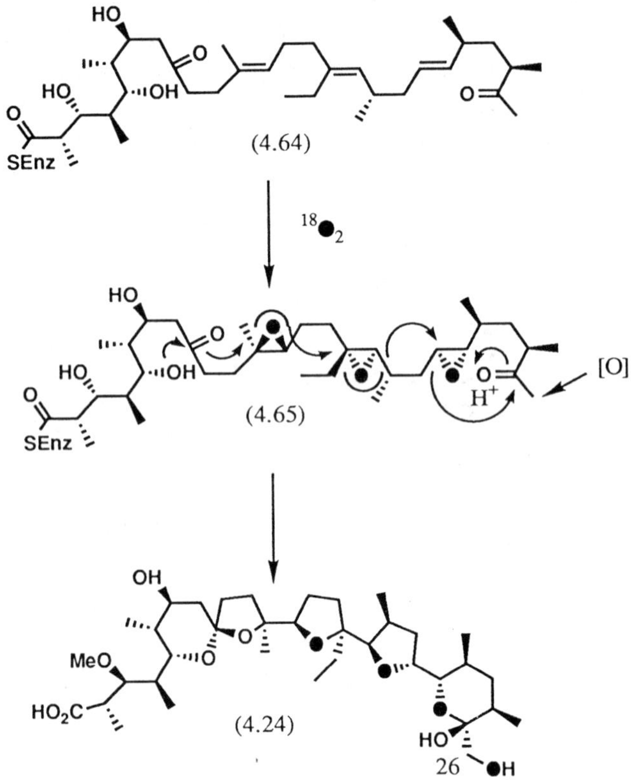

(4.64)

(4.65)

(4.24)

A great deal of effort has gone into trying to establish the integrity of intermediate polyenes and polypoxides, particularly in the monensin-A system. For example, partial fragments of the triene (4.64) and the entire triene itself have been synthesized and activated as their *N*-acylcysteamine thioesters, then introduced into whole cells of *Streptomyces cinnamonensis*, but as yet intact incorporations have been unsuccessful. Problems exist in the transport of such compounds across cell membranes, and those species that do enter the cell are efficiently degraded [60].

4.4.10 Polyether stereochemistry

In view of the stereochemical homologies already discussed for the macrolide (see section 4.4.5) and ansamycin (see section 4.4.7) antibiotics it is perhaps not surprising that there also exist significant stereochemical correlations within the polyether antibiotics. These correlations have been formalized by

Cane, Celmer and Westley who derived two prototypes (APPA and PAPA) that summarize the stereochemistry of more than thirty of the polyether antibiotics [61]. The designations APPA and PAPA relate to common combinations of acetate and propionate subunits which are utilized in the first four condensations during polyether biosynthesis. The prototypes summarize the structure and stereochemistry of the putative polyenes that are believed to be precursors of polyepoxide intermediates as discussed above. For example it can be seen that the structure of the putative triene precursor to monensin-A (4.64) is accommodated within the APPA prototype. It is noteworthy too that there exist areas of structural homology between the prototypes as indicated by the bracketed segments. Again we see evidence of ubiquitous regions of controlled stereochemistry in moving from one polyketide synthase to the next.

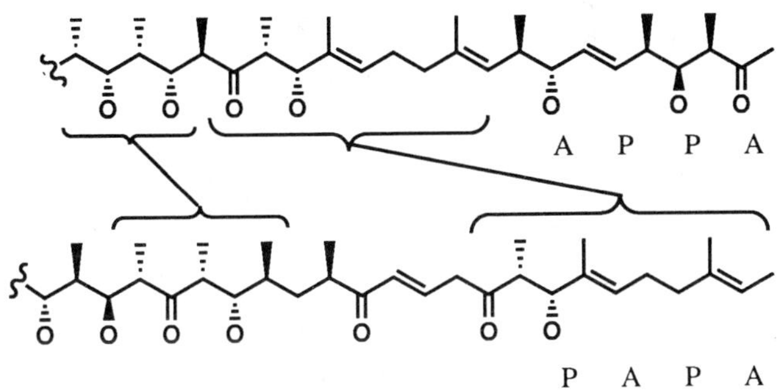

APPA prototype

PAPA prototype

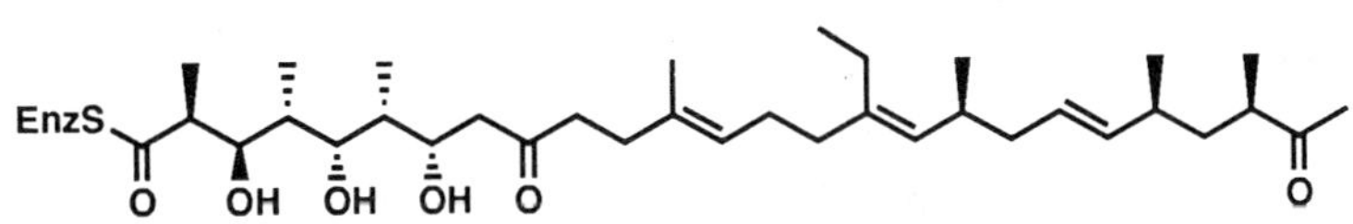

monensin-A polyene (4.64)

4.5 STRUCTURAL CORRELATIONS BETWEEN POLYKETIDE CLASSES

Structural similarities between the macrolide and ansamycin antibiotics have already been highlighted (see section 4.4.7). An analysis comparing the structure and stereochemistry of the polyether and macrolide antibiotics demonstrates too that there are substantial correlations between these classes. The most striking example can be seen when the aglycone of the 16-membered macrolide mycinamicin IV (4.52) is compared with the putative

triene (4.66)

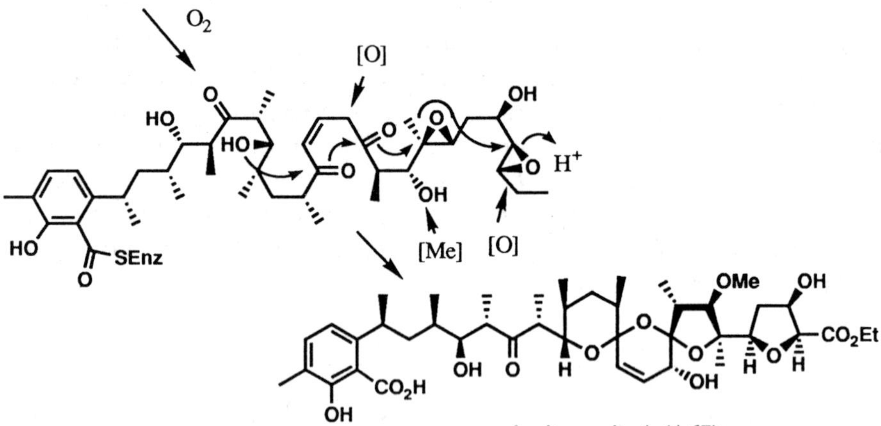

noboritomycin-A (4.67)

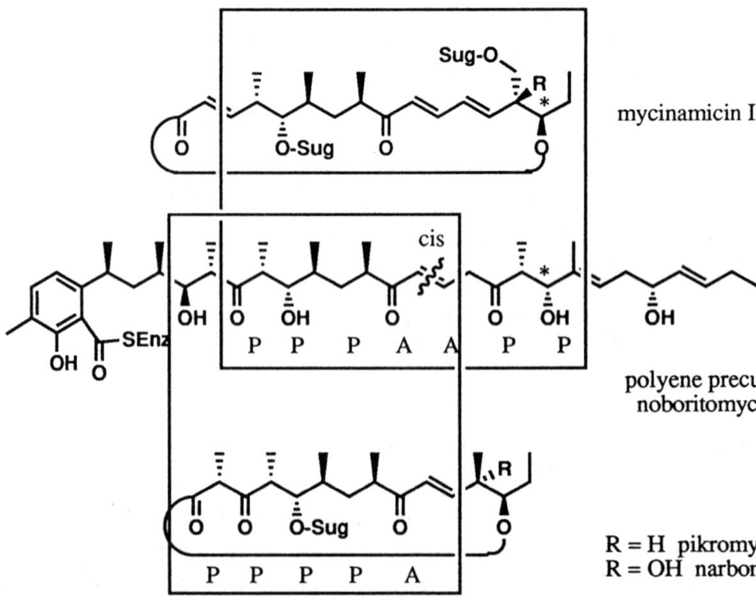

mycinamicin IV (4.52)

polyene precursor (4.66) to noboritomycin A

R = H pikromycin (4.68)
R = OH narbomycin (4.69)

triene precursor (4.66) to the *bis*dispiroketal polyether noboritomycin-A (4.67) [62]. There exists a seven-subunit sequence derived from propionate (P) and acetate (A), PPPAAPP, common to both sets of metabolites, over which the absolute stereochemistry of the oxygen and alkyl functionality matches at five of six asymmetric centres. Correlations are also found between the 12- and 14-membered macrolides and this polyene precursor. For the aglycones of pikromycin (4.68) and narbomycin (4.69), all five asymmetric centres over a common five-subunit sequence, (PPPPA) match, and there exists a common α,β-unsaturated functionality between the macrolides and the noboritomycin triene (4.66) [63].

The noboritomycin triene (4.66) appears to emerge as a focal point with structural features common to many of the Actinomycete metabolites and also to some marine polyketides (see section 5.5). It is tempting to speculate that the noboritomycin-A synthase is expressed by a particularly stable gene cluster which has resisted substantial evolutionary modifications. This triene may be the closest representation that we can deduce for the product of a primordial polyketide synthase.

4.6 MISCELLANEOUS MACROLIDES

There are many miscellaneous macrolides which do not fall neatly into obvious structural classes. It may be that some of these compounds are representatives of a much larger class of metabolites awaiting identification, and clearly a stimulus is provided for the continuation of screening programmes, particularly when compounds emerge that have unique and desirable biological activities. Interesting examples in this regard are FK 506 (4.70), a product of *Streptomyces tsukubaensis* [64], and a related metabolite, rapamycin (4.71), from *Streptomyces hygroscopicus* [65]. These compounds are currently attracting a great deal of interest owing to their powerful immunosuppressant activity [66]. They possess some interesting structural features such as a cyclohexyl ring system and, in FK 506, a pendent propenyl group which may derive from a C_5 pentanoic acid subunit? Other structural features of biosynthetic interest are the pyridinium ring and the α-ketoamide functionality, both of which play a key role in the biological activity of these metabolites.

Nargenicin-A (4.72) from *Nocardia argentinensis* possesses an unusual fused ring system and its biosynthesis has been investigated in detail. The evidence from isotopic labelling studies is consistent with a polyene precursor such as (4.73) as an enzyme-bound intermediate, which may undergo an intramolecular Diels–Alder type cycloaddition to set up the octalin ring system [67]. A partially elaborated C_6 fragment (4.74), representing an intermediate after the first condensation and reduction of polyketide assembly, was synthesized and activated as its *N*-acetylcysteamine thioester and incorporated intact into the nargenicin framework [68]. This fragment is identical to (4.48) which was incorporated into erythromycin-A (4.23) (see section 4.4.4), and clearly there is a similarity in the early stages of polyketide assembly.

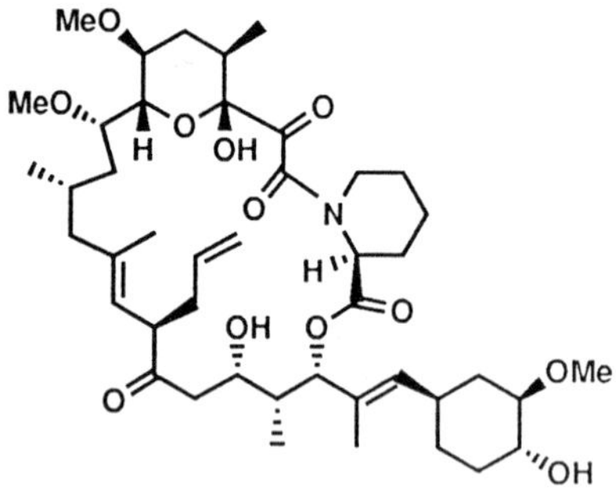

FK 506 (4.70)

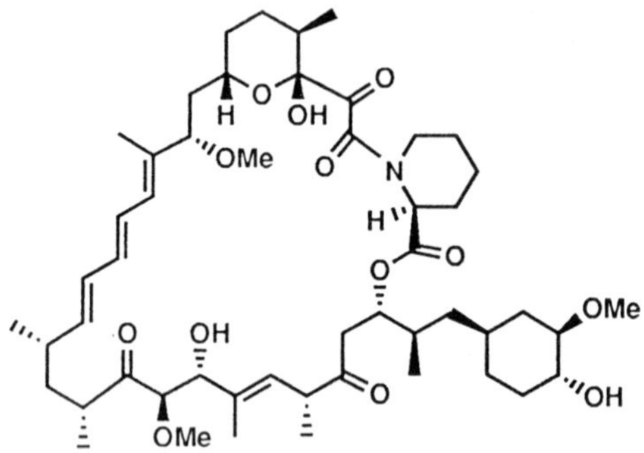

rapamycin (4.71)

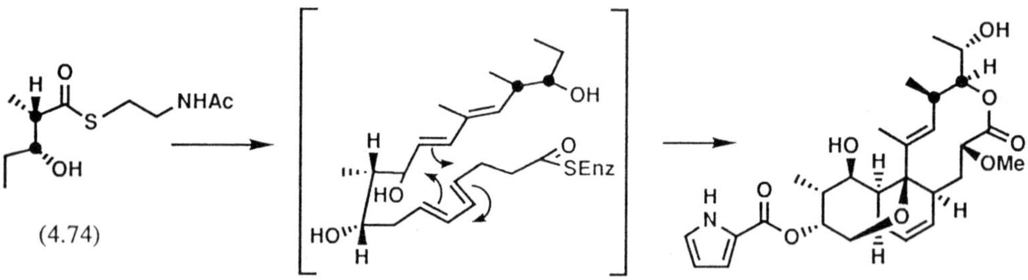

(4.74)

(4.73)

nargenicin-A (4.72)

Angiolam-A (4.75) from *Angiococcus disciformis* has a 19-membered ring system derived from two distinct polyketide assemblies [69]. It is constructed largely from acetate and propionate, and interestingly one of the polyketide chains is initiated from an L-alanine starter unit, although the α-carbon undergoes a configurational inversion during incorporation into the antibiotic. The structure and stereochemistry along the backbone of angiolam-A (4.75) from C-12 to C-19 is similar to the starter sequence of the APPA prototype which summarizes the structure of many polyether antibiotics. Whether these limited correlations are biogenetically significant is difficult to say with confidence. However, in the general sense, it is unlikely that the producing synthases of these macrolides are unrelated to those of the other higher polyketides.

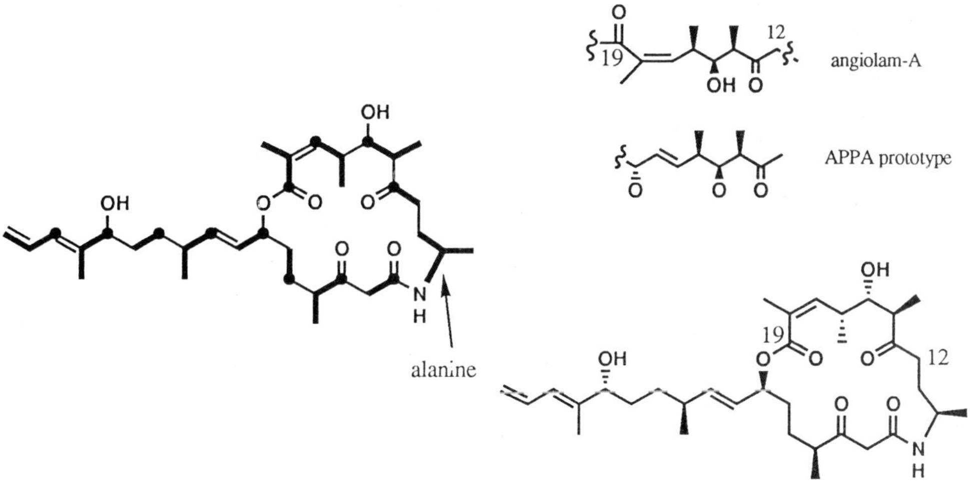

angiolam-A (4.75)

4.7 MICROBIAL ALKALOIDS

Alkaloids are normally associated with plants and they generally derive from amino acids, although we have already seen that there are a few plant alkaloids, exemplified by coniine and pinidine (see section 3.5.3), which have a polyketide origin. Not unsurprisingly perhaps there also exist a number of polyketide alkaloids from *Streptomyces* spp. Cycloheximide (4.11), which has already been highlighted (see section 4.1), can be placed in this category. The biosynthesis of nigrifactin (4.79) from *Streptomyces nigrifaciens* has been studied in some detail and it is derived from the folding of a hexatide precursor [70]. The double bonds appear to be introduced towards the end of the pathway, as evidenced by the efficient incorporation of (4.76), (4.77) and (4.78) into nigrifactin. Other *Streptomyces* alkaloids include latumcidin (4.80), pyrindicine (4.81), cyclizidine (4.82) and indolizomycin (4.83). Latumcidin (4.80) is derived from five acetate subunits [71] and pyrindicine (4.81) and from five acetates and a terminal propionate subunit [72]. The indolizidine alkaloid cyclizidine (4.82) has been shown to originate from a combination of acetate and propionate subunits, and the terminal cyclopropyl ring system is derived from the three carbons of the propionate starter unit [73]. Indolizomycin (4.83) is a structurally related metabolite which almost certainly has a similar mode of assembly and the details of its biosynthesis will prove interesting but await elucidation.

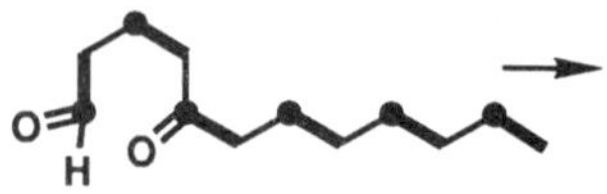

(4.76)

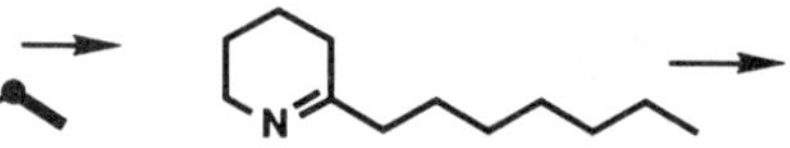

(4.77)

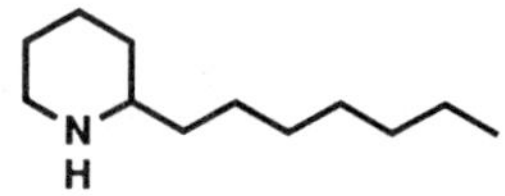

(4.78)

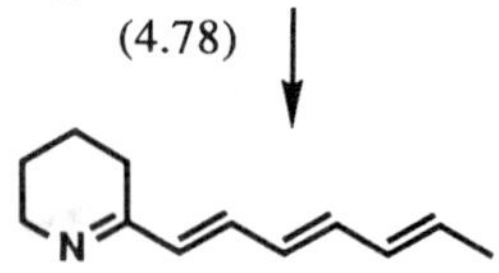

(4.79)

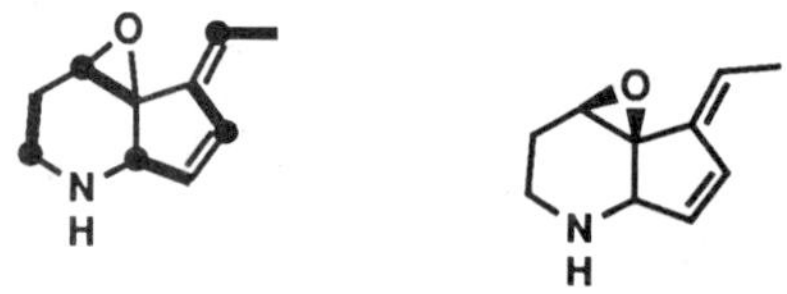

(4.80)

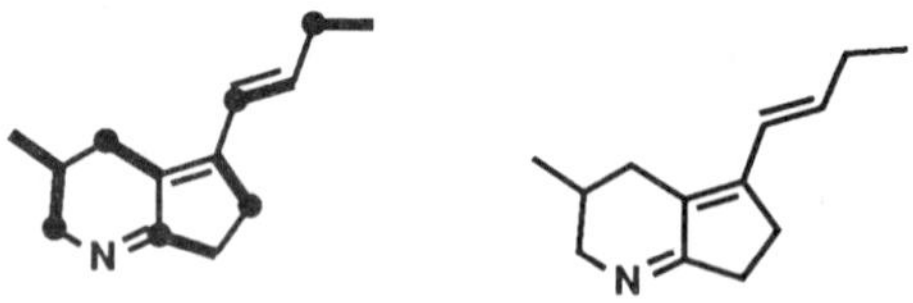

(4.81)

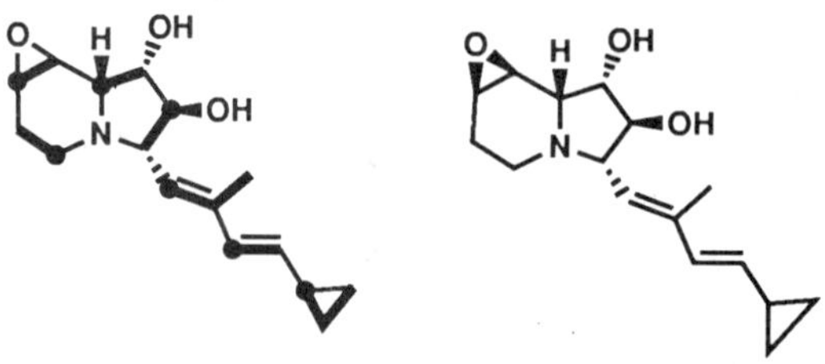

(4.82)

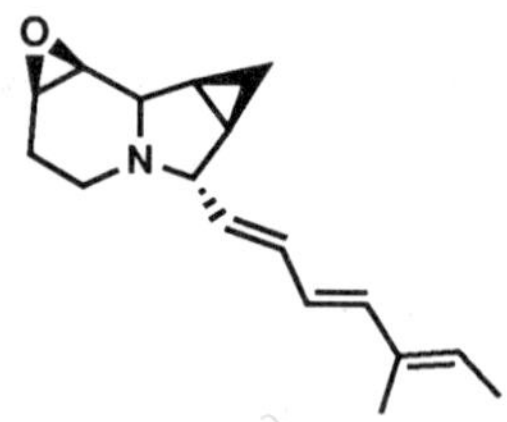

(4.83)

4.8 GENETIC ANALYSIS

4.8.1 Genetic homology

This chapter has attempted to highlight the existence of a structurally diverse range of polyketides, particularly from Actinomycetes. Despite this diversity there are clearly strong structural and in particular stereochemical relationships to be seen between many of these metabolites and this suggests that they have originated, during the evolutionary process, by a mechanism whereby genetic material from one synthase can contribute to a new set of polyketide synthase genes, such that a novel polyketide can emerge. In fact, the generation of hybrid antibiotics by recombinant techniques (see section 4.3) may be an accelerated model for the true evolutionary process. It is irrational to assume that polyketide synthases of such complexity, and which operate by comparable processes to generate structures which are often similiar, could have evolved in isolation from each other.

Consistent with this hypothesis is the identification of a substantial genetic homology between various polyketide producers, particularly from *Streptomyces spp*. [74]. The *act* I gene from *S. coelicolor* believed to code the condensing enzyme in actinorhodin (4.20) biosynthesis has been used as a hybridization probe to select homologous genetic material from DNA restriction digests of more than twenty other *Streptomyces* strains. Varing degrees of hybridization were detected in fourteen of eighteen known polyketide producers encompassing the tetracyclines, the macrolides, the polyethers and milbemycins as well as strong hydridization to other isochromanequinone antibiotics (granaticin, medermycin and kalafungin), the class to which actinorhodin itself belongs. A few strains (e.g. erythromycin, oxytetracycline) showed no hybridization. Interestingly, some *Streptomyces* strains that were not known to express polyketides showed some homology and these strains may have a cryptic capacity for polyketide production. Although not unanimous, these results demonstrate a strong correlation between antibiotic production and partially homologous DNA sequences and clearly the expression of related polyketide synthases is implied.

4.8.2 The structure of polyketide synthases

A body of information is beginning to accumulate on the structure of different polyketide synthases (PKSs) from various sources. The sequence analysis of the gene coding for 6-methylsalicylic synthase from the fungi *Penicillium patulum* [75], demonstrates that the protein is a single multifunctional protein consistent with the earlier protein purification analysis (see section 3.1.1) and must be considered a classic Type I PKS. Chalcone and resveratrol synthases from plants are also single proteins [76] and uniquely among the polyketide synthases do not possess a 4′-phosphopantetheinyl thiol, but processes coenzyme-A esters directly (see section 3.5.5).

A major effort in microbiology at present is focussed on the genetic analysis of PKS genes from bacterial systems and in particular from the commercially important Actinomycete antibiotic producers [77,78]. The system studied in the greatest detail is the isochromanequinone antibiotic actinorhodin (4.20) from *Streptomyces coelicolor*. Six *act* genes have been identified, located adjacent to each other in a cluster, which are believed to code for individual enzymes involved in polyketide assembly. For example the *act* I probably codes the condensing enzyme of the PKS as evidenced by its hybridization to other polyketide condensing enzymes (see section 4.8.1). The *act* III gene has been sequenced and the deduced protein sequence displays considerable homology to β-keto-reductase enzymes from fatty acid and other polyketide synthases [79]. It can be deduced therefore that the *act*

III gene codes the β-keto-reductase activity of the PKS. Complementary studies on several other aromatic polyketides, such as granaticin (4.19) from *S. violaceruber* [80] and the antitumour antibiotic tetracenomycin C from *S. glaucesences* [81] are providing a body of evidence which demonstrates that these PKSs are expressed by gene clusters where each gene is coding a discrete enzyme involved in polyketide assembly. One gene from each of these clusters would appear to code for an ACP. These are therefore Type II PKSs. In contrast however to the PKSs of these aromatic antibiotics from Actinomycetes preliminary investigations suggest that the higher polyketides such as the polyether and macrolide antibiotics have a Type I PKS. This is discussed in the next section (4.8.3).

4.8.3 Erythromycin and the module hypothesis

Analysis of the PKS genes of *Saccharopolyspora erythraea* suggests that the synthase responsible for the production of the erythromycin aglycone, 6-deoxyerythronolide-B (4.39), is composed of three multifunctional proteins rather than a series of discrete enzymes. An analogy can be drawn with the multifunctional Type I rather than the dissociated Type II fatty acid synthases (see section 1.2). For example, a 10 kb fragment of the DNA from *S. erythraea* has been identified and partially sequenced [82] and there is considerable homology between the deduced protein sequence and that from enzymes of fatty acid synthases from other sources. More particularly however, the code for *two sets* of four activities are apparent in the fragment i.e. there is a *repeat* sequence coding for a condensing activity (β-ketoacyl synthase) (KS), an acyltransferase (AT), a β-keto-reductase (KR) and then an acyl carrier protein (ACP) as illustrated in Fig. 4.2. At the terminus of this fragment there is also an acyl transferase or thioesterase (TE) which may be responsible for off-loading or cyclizing the assembled aglycone of the erythromycins. All of these nine activities are expressed in a single open reading frame (ORF-3) of 9.5 kb which implies that they are present on a multifunctional peptide rather than on individual proteins.

Additionally two other ORFs (ORF-1 and ORF-2) have been identified upstream of ORF-3 [83]. Again sequence analysis suggests that the resultant proteins coded by these ORFs each possess an ordered sequence of activities responsible for the partial assembly of 6-deoxyerythronolide-B as shown in Fig. 4.2. the genetic analysis of the erythromycin PKS from *S. erythraea* raises some interesting questions regarding the organization and control of higher polyketide synthases and provides evidence for a *modular* assembly of the polyketide aglycones. We have already seen that biosynthetic experiments support a processive construction of the erythromycin and other macrolide aglycones, where the reduction status of each β-carbonyl is adjusted prior to the subsequent condensation (see section 4.4.4). The

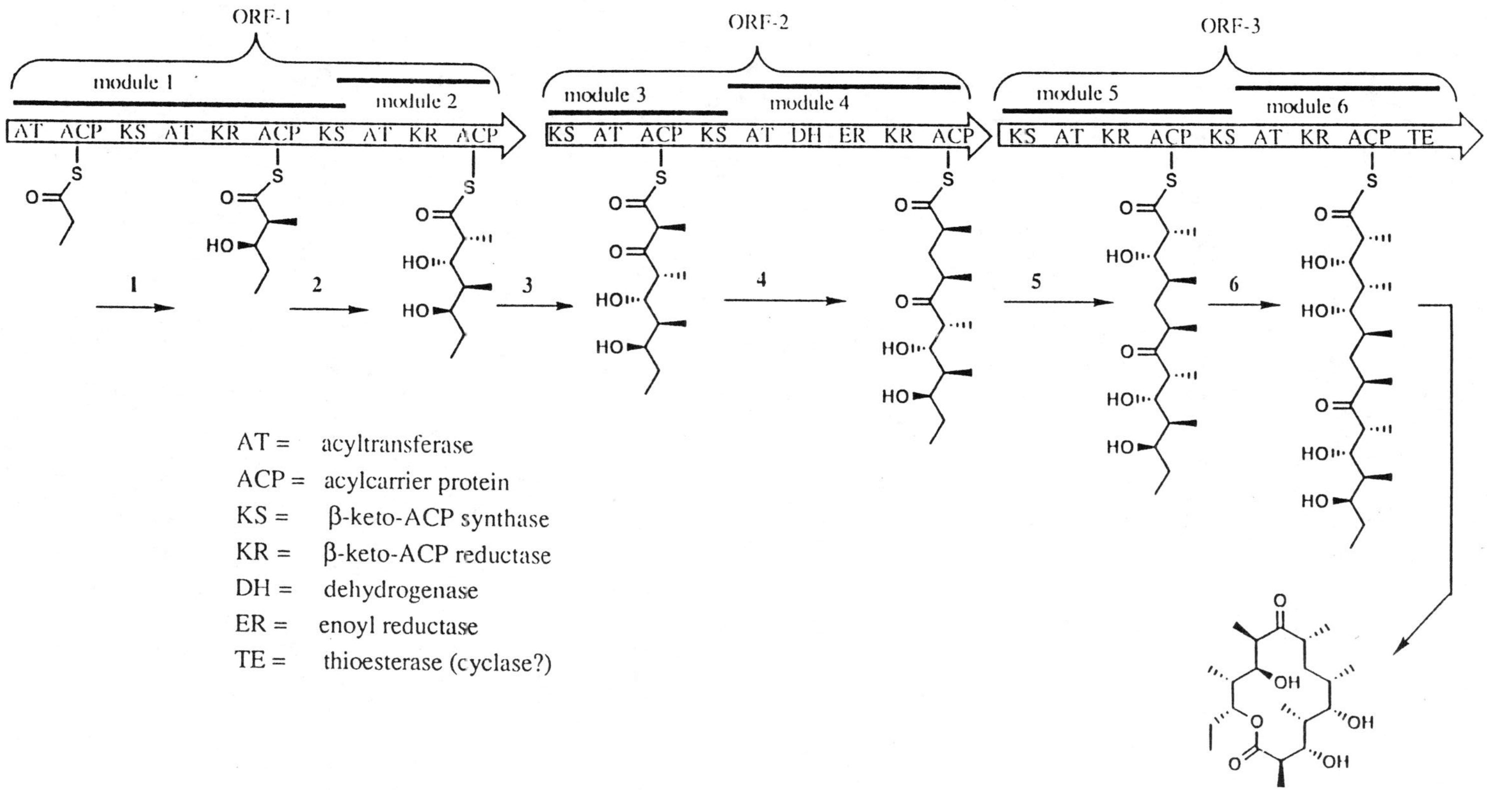

6-deoxyerythronolide-B (4.39)

Fig. 4.2 — The modular assembly of 6-deoxyerythronolide-B.

genetic analysis reinforces this conclusion. Each of the three multifunctional proteins coded by ORFs 1-3 could mediate two of the six 'propionate' condensation cycles (modules) required for 6-deoxyerythronolide-B (4.39) assembly. There are in total *seven* ACPs expressed over the three ORFs which raises the question of *separate* ACPs accepting the propionate starter unit and processing the developing acyl chains through the six condensations. By comparison, during fatty acid biosynthesis a *single* ACP transports the developing acyl moiety around all the various activities of the fatty synthase (see section 1.1). In view of the evidence from biosynthetic and genetic sources, a modular hypothesis can be rationalized for the construction of the higher polyketide metabolites [82]. In its extremity, each condensation/reduction cycle is mediated by a single ACP and all the enzymes required to adjust the β-keto-functionality to the necessary reduction status for that cycle, are arranged in sequence within the module as illustrated in Fig. 4.2.

The structural similarities which exist between polyketide groups, such as the macrolide-polyether (see section 4.5) and the macrolide-ansamycin (see section 4.4.7) homologies, may therefore arise from the expression of similar constituent multifunctional proteins mediating the assembly of common modules in different polyketide synthases. Also, those homologous regions between the APPA and PAPA prototypes for the polyether antibiotics (see section 4.4.10) could be the result of the expression of common modular sequences within the PKS at the genetic level. The evidence, although limited at present, is consistent with an evolutionary mechanism whereby Actinomycetes can 'mix and match' their biosynthetic genes to some extent in order to generate new polyketide synthases which retain structural features of the their originators.

REFERENCES

[1] *Encyclopedia of Antibiotics*, 2nd Edn. Ed. J. S. Glasby; John Wiley and Sons, 1978.

[2] J. R. D. McCormick, U. Hirsch, N. O. Sjolander and A. P. Doerschuk, Cosynthesis of tetracyclines by pairs of *Streptomyces aureofaciens* mutants. *J. Am. Chem. Soc.*, 1960, **82**, 5006.

[3] J. R. D. McCormick, U. H. Joachim, E. R. Jensens, S. Johnson and N. O. Sjolander, Biosynthesis of tetracyclines VIII. 4-Hydroxy-6-methyl-pretetramide, an intermediate accumulated by a blocked mutant of *Streptomyces aureofaciens*. *J. Am. Chem. Soc.*, 1965, **87**, 1793.

[4] J. R. D. McCormick, E. R. Jensen, N. H. Arnold, H. S. Corey, U. H. Joachim, S. Johnson, P. A. Miller and N. O. Sjolander, Biosynthesis of the tetracyclines XI. The methylanthrone analogue of protetrone. *J. Am. Chem. Soc.*, 1968, **90**, 7127.

[5] J. R. D. McCormick and E. R. Jensen, Biosynthesis of tetracyclines. X. Protetrone. *J. Am. Chem. Soc.*, 1968, **90**, 7126.

[6] R. Thomas and D. J. Williams, Oxytetracycline biosynthesis. Origin of the carboxamide substituent. *J. Chem. Soc. Chem. Commun.*, 1983, 677.

[7] P. W. Jeffs and D. McWilliams, Carbon-13 nuclear magnetic resonance study of the biosynthesis of cycloheximide. Stereospecific incorporation of malonate into the glutarimide ring, *J. Am. Chem. Soc.*, 1981, **103**, 6185; H. Shimada, H. Noguchi, Y. Iitaka and U. Sankawa, Biosynthesis of cycloheximide. Stereospecific incorporation of [1,2,3-$^{13}C_3$] malonate. *Heterocycles*, 1981, **15**, 1141.

[8] V. Behal, M. Podojil, Z. Hostalek, Z. Vanek and F. Lynen, Regulation of biosynthesis of excessive metabolites XVI. Origin of the terminal group of tetracyclines. *Folia Microbiol.*, 1974, **16**, 146.

[9] A. F. de Jesus, W. E. Hull, P. S. Steyn, F. R. van Heerden and R. Vleggaar, Biosynthesis of viridicatumtoxin, a mycotoxin from *Penicillium expansum*. *J. Chem. Soc. Chem. Commun.*, 1982, 902.

[10] R. C. Paulick, M. L. Casey and H. W. Whitlock, A ^{13}C-nuclear magnetic resonance study of the biosynthesis of daunomycin from $^{13}CH_3{}^{13}CO_2Na$, *J. Am., Chem. Soc.*, 1976, **98**, 3370.

[11] W. D. Ollis, I. O. Sutherland, R. C. Codner, J. J. Gordon and G. A. Miller, The incorporation of propionate in the biosynthesis of the ε-pyrromycinone (rutilantinone). *Proc. Chem. Soc., London*, 1960, 347.

[12] C. P. Gorst-Allman, B. M. A. Rudd, C.-J. Chang and H., G. Floss, Biosynthesis of actinorhodin. Determination of the point of dimerisation *J. Org. Chem.*, 1981, **46**, 455.

[13] D. A. Hopwood, F. Malpartida, H. M. Kieser, H. Ikeda, J. Duncan, I. Fujii, B. A. M. Rudd, H. G. Floss and S. Omura, Production of 'hybrid' antibiotics by genetic engineering. *Nature*, 1985, **314**, 642.

[14] H. G. Floss, Hybrid antibiotics — the contribution of the new gene combinations, *TIBTECH*, 1987, **5**, 111.

[15] K. Gerzon, E. H. Flynn, M. V. Sigal, P. F. Wiley, R. Monahan and U. C. Quarck, Erythromycin VIII. Structure of dihydroerythronolide. *J. Am. Chem. Soc.*, 1956, **78**, 6398.

[16] R. B. Woodward, Struktur und biogenese deer makrolide. *Angew. Chem.*, 1957, **69**, 50.

[17] S. M. Friedman, T. Kaneda and J. W. Corcoran, A comparison of 2-methylmalonate and propionate as precursors of the C_{21} branched chain lactone in erythromycin. *J. Biol. Chem.*, 1964, **239**, 2386.

[18] S. Omura, K. Tsuzuki, A. Nakagawa and G. Lukacs, Biosynthesis and origin of carbons 3 and 4 of the leucomycin aglycone. *J. Antibiotics*, 1983, **36**, 611.

[19] D. M. Ashworth and J. A. Robinson, Biosynthesis of the macrotetrolide antibiotics. An investigation using carbon-13 and oxygen-18

labelled acetate and propionate. *J. Chem. Soc. Chem. Commun.*, 1983, 1327.

[20] Z. M. Spavold and J. A. Robinson, Nonactin biosynthesis. On the role of (6*R*,8*R*)- and (6*S*,8*S*)-2-methyl-6,8-dihydroxynon-2*E*-enoic acids in the formation of nonactic acid. *J. Chem. Soc. Chem. Commun.*, 1988, 4.

[21] R. J. Parry and R. Mafoti, Biosynthesis of coronatine, a novel polyketide. *J. Am. Chem. Soc.*, 1986, **108**, 4681.

[22] A. R. Hunaiti and P. E. Kolattukudy, Isolation and characterisation of an acyl-coenzyme carboxylase from the erythromycin producing *Streptomyces erythreus. Arch. Biochem. Biophys.*, 1982, **216**, 362.

[23] C. R. Hutchinson, M. M. Sherman, A., G. McInnes, J. A. Walter and J. C. Vederas, Biosynthesis of macrolides. 6. Mechanism of stereocontrol during formation of lasalocid A. *J. Am. Chem. Soc.,* 1981, **103**, 5956.

[24] G. R. Sood, J. A. Robinson and A. A. Ajaz, Biosynthesis of the polyether antibiotic monensin-A. Incorporation of [2-$^{2}H_{2}$]-, (*R*)-[2-$^{2}H_{1}$] and (*S*)[2-$^{2}H_{1}$]-propionate. *J. Chem. Soc. Chem. Commun.*, 1984, 1421.

[25] D. E. Cane, T.-C. Laing, P. B. Taylor, C. Chang and C.-C. Yang, Macrolide biosynthesis 3. Stereochemistry of the chain elongation steps of erythromycin biosynthesis. *J. Am. Chem. Soc.*, 1986, **108**, 4957.

[26] M. M. Sherman and C. R. Hutchinson, Biosynthesis of lasalocid-A. Biochemical mechanism for the assembly of the carbon framework. *Biochemistry*, 1987, **26**, 438.

[27] D. E. Cane, H. Hasler, P. B. Taylor and T.-C. Liang, Origin of the carbon skeleton and oxygen atoms of the erythromycins. *Tetrahedron*, 1983, **39**, 3449.

[28] D. E. Dorman, J. W. Pascal, W. M. Nakatsukasa, L. L. Huckstep and N. Neuss, Biosynthesis of narasin, a new polyether ionophore from the fermentation of *Streptomyces aureofaciens. Helv. Chem. Acta*, 1976, **59**, 2625.

[29] K. A. Reynolds, D. O'Hagan, D. Gani and J. A. Robinson, Butyrate metabolism in Streptomycetes. Characterisation of an intramolecular vicinal rearrangement linking isobutyrate and butyrate in *Streptomyces cinnamonensis. J. Chem. Soc. Perkin Trans.*, I, 1988, 3195.

[30] S. Omura, K. Tsuzuki, Y. Tanaka, H. Sakakibara, M. Aizawa and G. Lukacs, Valine as a precursor of the *n*-butyrate unit in the biosynthesis of macrolide aglycone. *J. Antibiotics*, 1983, **36** 614.

[31] M. M. Sherman, S. Yue and C. R. Hutchinson, Biosynthesis of lasalocid-A. Metabolic interrelationships of carboxylic acid precursors and polyether antibiotics. *J. Antibiotics*, 1986, **39**, 1135.

[32] G. Brendelberger, J. Retey, D. M. Ashworth, K., Reynolds, F. Willenbrock and J. A. Robinson, The enzymatic interconversion of isobutyryl and *n*-butyrycarba(dethia)-coenzyme-A: a coenzyme-B_{12} dependent carbon skeleton rearrangement. *Angew. Chim. Int. Ed. Engl.*, 1988, **27**, 1089.

[33] J. R. Martin and W. Rosenbrook, Studies on the biosynthesis of the erythromycins. II. Isolation and structure elucidation of a biosynthetic intermediate, 6-deoxyerythronolide-B. *Biochemistry*, 1967, **6**, 435.

[34] T. Furumai and M. Suzuki, Biosynthesis of basic 16-membered macrolide antibiotics, platenomycins III. Production, isolation and structures of platenolides I and II. *J. Antibiotics*, 1975, **28**, 783.

[35] K. Kinoshita, S. Takenaka and M. Hayashi, Isolation of intermediates in the biosynthesis of mycinamicins. *J. Chem. Soc. Chem. Commun.*, 1988, 943.

[36] M. L. B. Huber, J. W. Paschal, J. P. Leeds, H. A. Hirst, J. A. Wind, F. D. Miller and J. R. Turner, Branched chain fatty acids produced by mutants of *Streptomyces fradiae*, putative precursors of the lactone-ring of tylosin. *Anti. Microb. Ag. Chemother.*, 1990, **34**, 1535.

[37] S. Yue, J. S. Duncan, Y. Yamamoto and C. R. Hutchinson, Macrolide biosynthesis. Tylactone formation involves the processive addition of three carbon units. *J. Am. Chem. Soc.*, 1987, **109**, 1253.

[38] D. E. Cane and C.-C. Yang, Macrolide biosynthesis 4. Intact incorporation of a chain elongation intermediate into erythromycin. *J. Am. Chem. Soc.*, 1987, **109**, 1255.

[39] D. O'Hagan, J. A. Robinson and D. L. Turner, Biosynthesis of the macrolide antibiotic tylosin. Origin of the oxygen atoms of tylactone. *J. Chem. Soc. Chem. Commun.*, 1983, 1337.

[40] D. E. Cane, H. Hasler, P. B. Taylor and T.-C. Liang, Macrolide biosynthesis II. Origin of the carbon skeleton and oxygen atoms of the erythromycins. *Tetrahedron*, 1983, **39**, 3449.

[41] W. D. Celmer, Macrolide stereochemistry III. A configurational model for the macrolide antibiotics. *J. Am. Chem. Soc.*, 1965, **87**, 1801.

[42] J. Clardy, F. Finer-Moore, L. Weiler and D. C. Wiley, The structure of acumycin. A 16-membered ring macrolide antibiotic. *Tetrahedron*, 1981, **37**, Supplement No. 1, 91.

[43] G. Albers-Schonberg, B. H. Arison, J. C. Chabala, A. W. Douglas, P. Eskola, M. H. Fisher, A. Lusi, H. Mrozik, J. L. Smith and R. L. Tolman, Avermectins. Structure determination. *J. Am. Chem. Soc.*, 1981, **103**, 4216.

[44] Y. Takiguchi, M. Ono, S. Muramatsu, J. Ide, H. Mishima and M. Terao, Milbemycins, a new family of macrolide antibiotics. Fermentation, isolation and physiochemical properties of milbemycins D, E, F, G and H. *J. Antibiotics*, 1983, **36**, 502.

[45] M. Ono, M. Hiroshi, Y. Takiguchi and M. Terao, Milbemycins, a new family of macrolide antibiotics. Fermentation, isolation physicochemical properties and bioconversion of milbemycins J and K. *J. Antibiotics*, 1983, **36**, 509.

[46] D. E. Cane, T.-C. Liang, L. Kaplan, M. K. Nallin, M. D. Schulman, O. D. Hensens, A. W. Douglas and G. Albers-Schonberg, Biosynthetic origin of the carbon skeleton and oxygen atoms of the avermectins. *J. Am. Chem. Soc.*, 1983, **105**, 4110.

[47] O. Ghisalba, H. Fuhrer, W. J. Richter and S. Moss, A genetic approach to the biosynthesis of the rifamycin chromophore in *Norcadia mediterranei. J. Antibiotics*, 1981, **34**, 58.

[48] R. J. White, E. Martinelli, G. G. Gallo, G. Lancini and P. Beynon, Rifamycin biosynthesis studied with ^{13}C-enriched precursors and carbon magnetic resonance. *Nature*, 1973, **243**, 273.

[49] V. Prelog and W. Oppolzer, Ansamycine, eine neuartige Klasse von mikrobiellen Stoffwechselprodukten. *Helv. Chim. Acta*, 1973, **56**, 2279.

[50] W. Mechlinski, C. P. Schaffner, P. Ganis and G. Avitabile, Structure and absolute configuration of the polyene macrolide amphotericin B. *Tetrahedron Lett.*, 1970, 3873.

[51] P. H. Harrison, H. Noguchi and J. C. Vederas, Biosynthesis of polyene antibiotics: intact incorporation of ^{13}C-labelled octanoate into fungichromin by *Streptomyces cellulosae. J. Am. Chem. Soc.*, 1986, **108**, 3833.

[52] J. W. Westley, R. H. Evans Jr., T. Williams and A. Stempel, Structure of antibiotic X-537-A. *J. Chem. Soc. Chem. Commun.*, 1970, 71.

[53] J. W. Westley, J. F. Blount, R. H. Evans Jr., A. Stempel and J. Berger, Biosynthesis of lasalocid II. X-Ray analysis of a naturally occurring isomer of lasalocid-A *J. Antibiotics*, 1974, **27**, 597.

[54] C. R. Hutchinson, M. M. Sherman, J. C. Vederas and T. T. Nakashima, Biosynthesis of macrolides 5. Regiochemistry of the labelling of lasalocid-A by ^{13}C, ^{18}O- labelled precursors. *J. Am. Chem. Soc.*, 1981, 103, 5953.

[55] D. E. Cane, T.-C. Liang and H. Hasler, Polyether biosynthesis 2. Origin of the oxygen atoms of monensin-A. *J. Am. Chem. Soc.*, 1982, **104**, 7274.

[56] A. A. Ajaz, J. A. Robinson and D. L. Turner, Biosynthesis of the polyether ionophore antibiotic monensin-A. Assignment of the carbon-13 and proton NMR spectra of monensin-A by two dimensional spectroscopy. Incorporation of oxygen-18 labelled molecular oxygen. *J. Chem. Soc., Perkin Trans. 1,* 1987, 27.

[57] H. R. Tsou, S. Rajan, T. T. Chang, R. R. Fiala, G. W. Stockton and M.

W. Bullock, The utilisation of molecular oxygen during the biosynthesis of maduramicin. *J. Antibiotics*, 1987, **40**, 94.

[58] Z. Spavold, J. A. Robinson and D. L. Turner, Biosynthesis of the polyether antibiotic narasin. Origins of the oxygen atoms and the mechanism of ring formation. *Tetrahedron Lett.*, 1986, **27**, 3299.

[59] D. E. Cane and B. R. Hubbard, Polyether biosynthesis. Origin of the carbon skeleton and oxygen atoms of lenoremycin. *J. Am. Chem. Soc.*, 1987, **109**, 6533.

[60] D. S. Holmes, J. A. Sherringham, U. C. Dyer, S. T. Russell and J. A. Robinson, Synthesis of putative intermediates on the monensin biosynthetic pathway and incorporation experiments with the monensin producing organism. *Helv. Chim. Acta*, 1990, **73**, 239.

[61] D. E. Cane, W. D. Celmer and J. W. Westley, Unified stereochemical model of polyether antibiotic structure and biogenesis. *J. Am. Chem. Soc.*, 1983, **105**, 3594.

[62] D. O'Hagan, Structural and stereochemical homology between the macrolide and polyether antibiotics. *Tetrahedron*, 1988, **44**, 1691.

[63] D. O'Hagan, The polyether and macrolide antibiotics. Biogenetic analysis and structural correlations. *Nat. Prod. Reports*, 1989, **6**, 205.

[64] H. Tanaka, A, Kuroda, H., Marusawa, H. Hatanaka, T. Kino, T. Goto and M. Hashimoto, Structure of FK 506. A novel immunosuppressant isolated from *Streptomyces*. *J. Am. Chem. Soc.*, 1987, **109**, 5031.

[65] J. A. Findlay and L. Radics, On the chemistry and high field nuclear magnetic resonance spectroscopy of rapamycin. *Can. J. Chem.*, 1980, **58**, 579.

[66] M. W. Harding, A. Galat, D. E. Uehling and S. L. Schreiber, A receptor for the immunosuppressant FK 506, is a *cis-trans* peptidyl–prolyl isomerase. *Nature*, 1989, **341**, 758.

[67] D. E. Cane and C.-C. Yang, Biosynthetic origin of the carbon skeleton and oxygen atoms of nargenicin-A. *J. Am. Chem. Soc.*, 1984, **106**, 784.

[68] D. E. Cane and W. R. Ott, Macrolide biosynthesis 5. Intact incorporation of a chain elongation intermediate into nargenicin. *J. Am. Chem. Soc.*, 1988, **110**, 4840.

[69] W. Kohl, B. Witte, B. Kunze, V. Wray, D. Schomburg, H. Reichenbach and G. Hofle, Angiolam A-ein neues antibiotikum aus *Angiococcus disciformis* (Myxobacterales). *Liebigs Ann. Chem.*, 1985, 2088.

[70] T. Terashima, R. Idaka, Y. Kishi and T. Goto, Biosynthesis of nigrifactin, *J. Chem. Soc. Chem. Commun.*, 1973, 75.

[71] H. Seto, T. Sato and H. Yonehara, Utilisation of carbon-13–carbon-13 coupling in structural and biosynthetic studies. *J. Am. Chem. Soc.*, 1973, **95**, 8461.

[72] Y. Iwai, K. Kumano and S. Omura, Biosynthetic studies of microbial

alkaloid pyrindicine using carbon-13 labelled precursors. *Chem. Pharm. Bull.*, 1978, **26**, 736.

[73] F. J. Leeper, P. Padmanabhan, G. W. Kirby and G. N. Sheldrake, Biosynthesis of the indolizidine alkaloid cyclizidine. *J. Chem. Soc. Chem. Commun.*, 1987, 505.

[74] F. Malpartida, S. E. Hallam, H. M. Kieser, H. Motamedi, C. R. Hutchinson, M. J. Butler, D. A. Sugden, M. Warren, C. McKillop, C. R. Bailey, G. O. Humphreyes and D. A. Hopwood, Homology between Streptomyces genes coding for the synthesis of different polyketides used to clone antibiotic biosynthesis genes. *Nature*, 1987, **325**, 818.

[75] J. Beck, S. Ripka, A. Siegner, E. Schiltz and E. Schweizer, The multifunctional 6-methylsalicylic acid synthase gene of *Penicillium patulum*: its gene structure relative to that of other polyketide synthases. *Eur. J. Biochem.*, 1990, **192**, 487.

[76] T. Lanz, S. Tropf, F.-J. Marner, J. Schroder and G. Schroder, The role of cysteines in polyketide synthases. Site directed mutagenesis of resveratrol and chalcone synthases, two key enzymes in different plant specific pathways. *J. Biol. Chem.*, 1991, **266**, 9971.

[77] D. A. Hopwood and D. H. Sherman, Molecular genetics of polyketides and its comparison to fatty acid biosynthesis. *A. Rev. Genet.*, 1990, **24**, 37.

[78] J. A. Robinson, Polyketide synthase complexes; their structure and function in antibiotic biosynthesis, *Phil. Trans. R. Soc. Lond. B.*, 1991, **332**, 107.

[79] S. E. Hallam, F. Malpartida and D. A. Hopwood, nucleotide sequence, transcription and deduced function of a gene involved in polyketide antibiotic synthesis in *Streptomyces coelicolor. Gene,* 1988, **74**, 305.

[80] D. H. Sherman, F. Malpartida, M. J. Bibb, H. M. Kieser, M. J. Bibb and D. A. Hopwood, Structure and deduced function of the granaticin producing polyketide synthase gene cluster of *Streptomyces violaceoruber* Tu22. *EMBO J.*, 1989, **8**, 2717.

[81] M. J. Bibb, S. Biro, H. Motamedi, J. Collins and C. R. Hutchinson, Analysis of the nucleotide sequence of the *Streptomyces glaucescens tcml* genes provides key information about the enzymology of polyketide antibiotic biosynthesis. *EMBO J.*, 1989, **8**, 2727.

[82] J. Cortes, S. F. Haydock, G. A. Roberts, D. J. Bevitt and P. F. Leadlay, An unusually large multifunctional polypeptide in the erythromycin producing polyketide synthase of *Saccharopolyspora erythraea. Nature* 1990, **348**, 176.

[83] S. Conadio, M. J. Staver, J. B. McAlpine, S. J. Swanson and L. Katz, Modular organization of the genes required for complex polyketide biosynthesis, *Science*, 1991, **252**, 675.

5

Marine polyketides

Marine organisms are increasingly being recognized as a rich source of biologically active metabolites, and consequently much effort is focused on identifying new metabolites from marine sources. Organisms in the aquatic environment encompass prokaryotes such as cyanobacteria and a whole spectrum of eukaryotic species, extending from unicellular protozoa and phytoplankton to multicellular organisms such as corals and sponges. Animals are represented by the marine invertebrates such as mollusks and gastropods, and it is not surprising that a great variety of metabolites have been identified from such a rich biological arena [1]. It is, however, notoriously difficult to establish the origin of specific metabolites, particularly when they are identified from organisms that filter-feed (e.g. sponges, brachiopods, etc.). Clearly these organisms can accumulate metabolites from micro-organisms such as plankton rather than synthesize them independently. The common symbiotic relationships which exist between the micro-organisms and higher organisms such as sponges and corals further confound the issue.

Biosynthetic investigations are difficult to carry out on marine organisms. Some marine bacteria have been isolated and cultured. The unicellular eukaryotes can also be cultured, often in the presence of antibiotics, to generate aseptic growth, and some interesting studies (e.g. on brevetoxins, see section 5.2) have been carried out in this way. For higher organisms, however, the low concentration of available organic nutrients in the sea has resulted in very slow rates of secondary metabolite synthesis, forcing prolonged biosynthetic experiments which are often difficult to sustain either *in situ* (e.g. on a coral reef) or in an aquarium. This is a challenging area of investigation in which the literature is only just developing [2].

Most classes of polyketides are represented from marine organisms. In Chapter 1 (see section 1.5.3) a few examples of long chain fatty acids, particularly from sponges, were highlighted. The intriguing and abundant prostanoid metabolites from soft corals have been discussed in some detail in Chapter 2 (see section 2.2.1). This chapter focuses largely on the structurally more complex polyketides secreted by marine organisms. There is limited biosynthetic evidence to draw upon and therefore the polyketide origin of many of the metabolites highlighted is assumed on the basis of structural analogy to terrestrial systems or hypotheses which have been proposed in the literature.

5.1 MARINE BACTERIA

Aplasmomycin (5.1) is an ionophore macrolide antibiotic isolated from strain SS-20 of *Streptomyces griseus* (Actinomycete) obtained from a shallow marine mud. The strain requires saline water to produce the antibiotic and therefore is an interesting example on the interface between a terrestrial and marine bacteria. Aplasmomycin is one of a rare group of natural products which contain boron, and it biosynthesis has been studied in detail [3]. It is a dimeric structure, with each monomeric unit constructed from a polyketide chain derived from acetate and branching methyl groups from methionine. An interesting feature of polyketide assembly is a C3 starter unit which is labelled efficiently by glycerol. Thioesters of phosphoenolpyruvate or phosphoglycerate are considered the most likely candidates for the true starter unit, although this is difficult to establish as appropriately labelled phosphates do not penetrate cell walls [3].

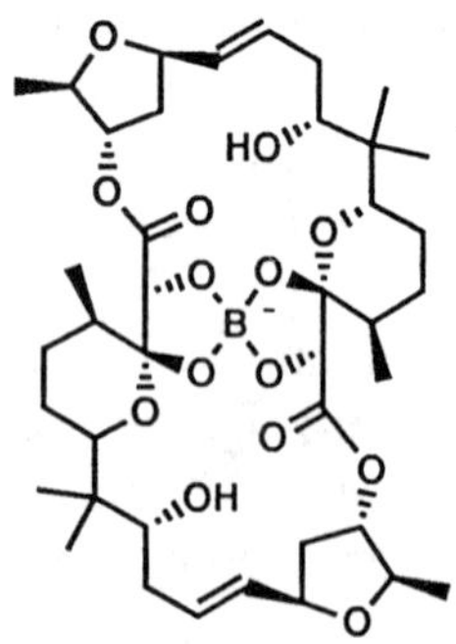

aplasmomycin (5.1)

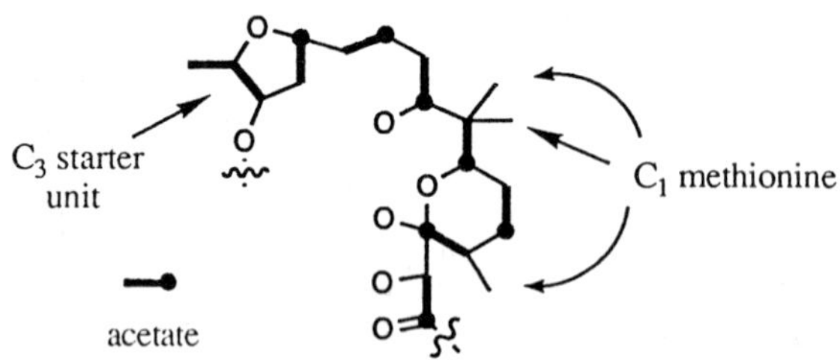

Some marine bacteria have been shown to produce aromatic polyketides. For example the rubrenoic acids A, B and C (5.2–5.4) from *Alteromonas rubra* [4] are relatively simple aromatic acids which almost certainly derive from the elaboration of unsaturated fatty acids. These compounds display bronchodilator activity. The more elaborate anthraquinone anitibiotics SS-228Y (5.5) and SS-228R (5.6) were isolated from a species of *Chainia* collected from shallow sea mud [5]. Their structures appear to be established as shown after a reassessment [6]. Interestingly, SS-228R can be accounted for by the rearrangement and dehydration of SS-228Y in light, a process which represents an alternative entry into linear tetracyclic systems (see sections 4.1 and 4.2). No biosynthetic studies have been carried out on the marine organism, although the origin of the structurally analogous aglycone of vineomycin-A_1 (5.7), an antitumor antibiotic from *Streptomyces matensis vineus* (terrestrial Actinomycete), has been investigated [7]. Labelling studies using acetate have demonstrated that this aglycone is derived from a decaketide with loss, through decarboxylation, of the terminal carboxylate group. Caution does have to be exercised when comparing biosynthetic pathways between different organisms; however, both are prokaryotic bacteria and an analogous biosynthesis appears likely.

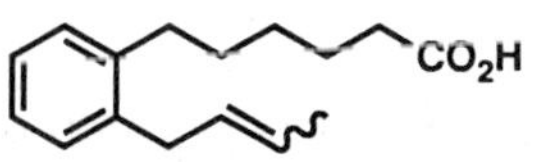

(5.2)

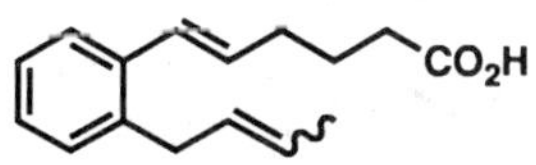

(5.3)

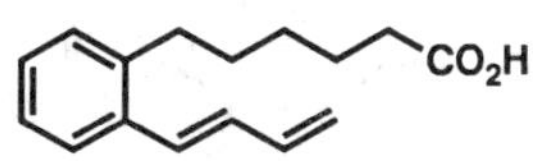

(5.4)

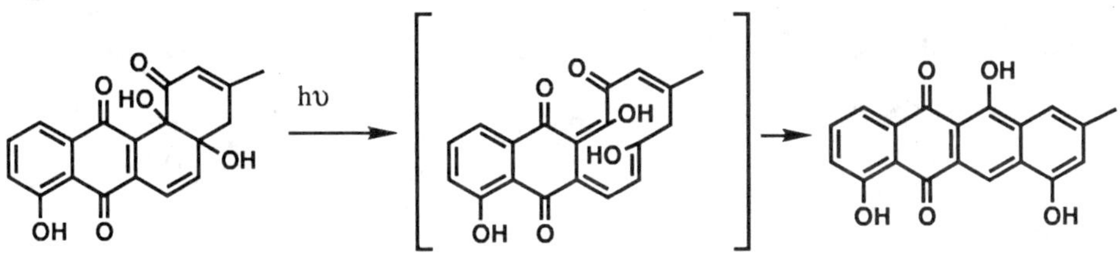

SS-228Y (5.5)

SS-228R (5.6)

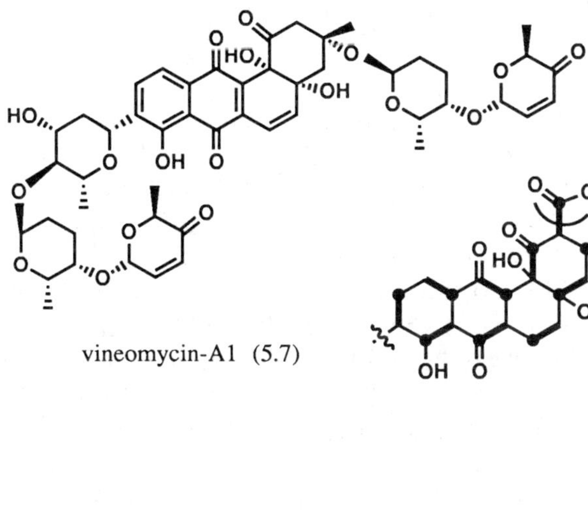

vineomycin-A1 (5.7)

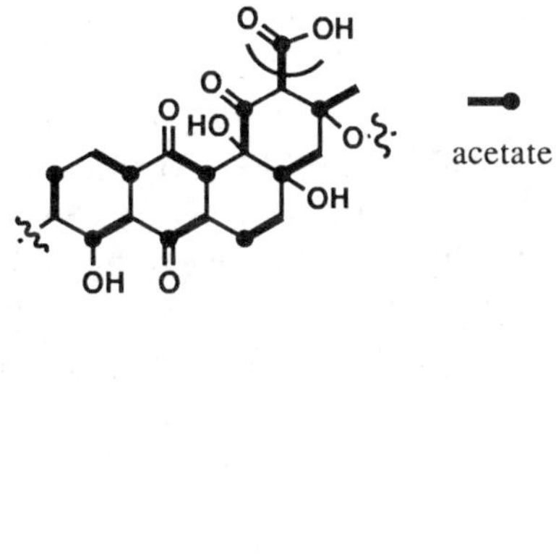

The cyanobacteria (blue-green algae) are responsible for a variety of complex secondary metabolites. Modified fatty acids are common, such as malyngic acid (5.8) [8] and (−)-(4*E*,7*S*)-7-methoxytetradec-4-enoic acid (5.9) from shallow water varieties of *Lyngbya majuscula*. Malygamide-A (5.10) was identified from a deep water variety of this bacteria and is an amide of (5.9), itself a constituent of the lipids in this variety [9]. The cyclodepsipeptide dolstatin-14 (5.11) displays antineoplastic activity and was isolated from the shell-less mollusk *Dolabella auricularia* [10]; however, dolstatin-14, or possibly its component parts, are believed to originate from blue-green algae. The structural similarity between (5.9) and the lipophilic region of dolstatin-14 (5.11) lends support to this hypothesis. The δ-lactone, malyngolide (5.12) is a further example of a modified fatty acid metabolite from *Lyngbya majuscula* [11].

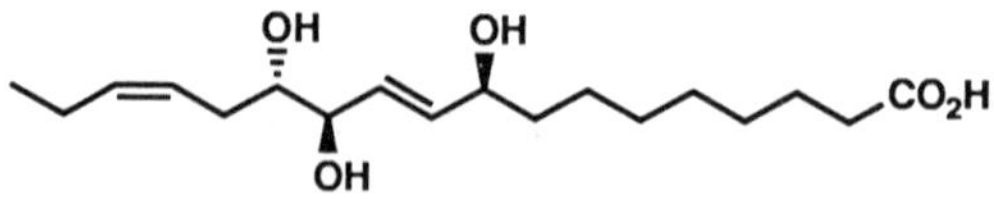

(5.8)

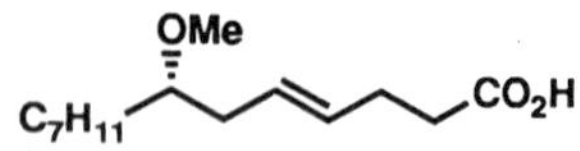

(5.9)

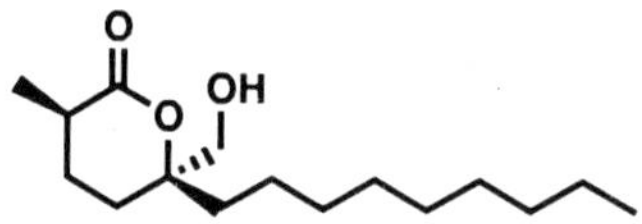

(5.12)

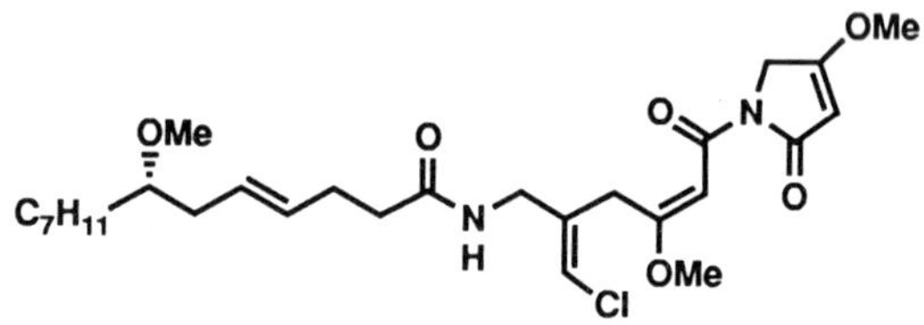

(5.10)

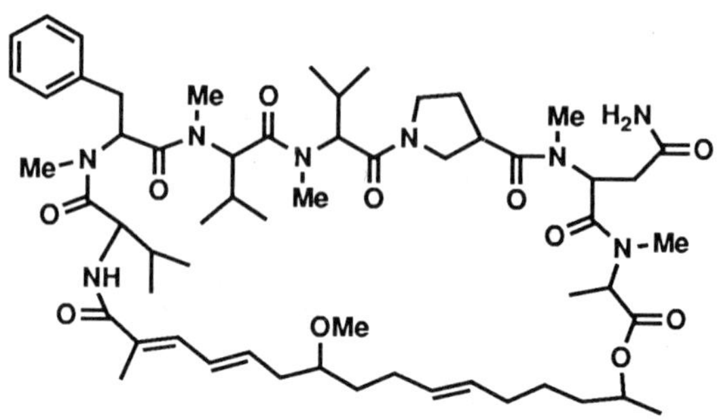

(5.11)

Among the most interesting polyketide structures isolated from cyanobacteria so far are the macrocyclic lactones scytophycins-A and -B (5.13 and 5.14) [12]. Scytophycins-A and -B were secreted from the cultured cells of *Scytonema pseudohofmanni*, a terrestrial blue-green algae. The terrestrial nature of the originating algae is interesting particularly in view of the

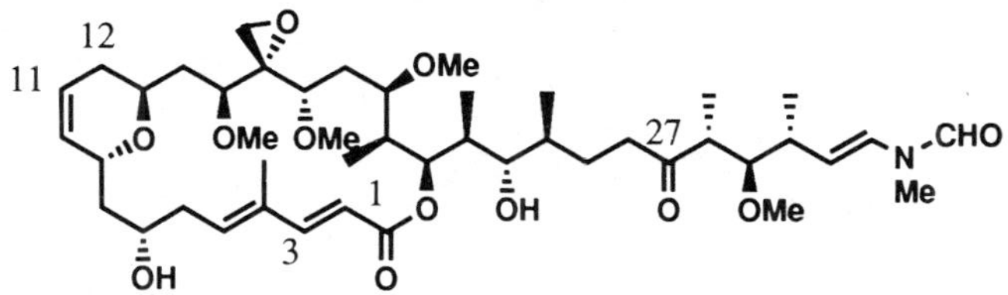

scytophycin-A (5.13)

(5.14) same as (5.13) except at C27:

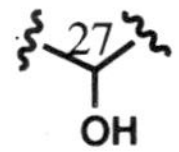

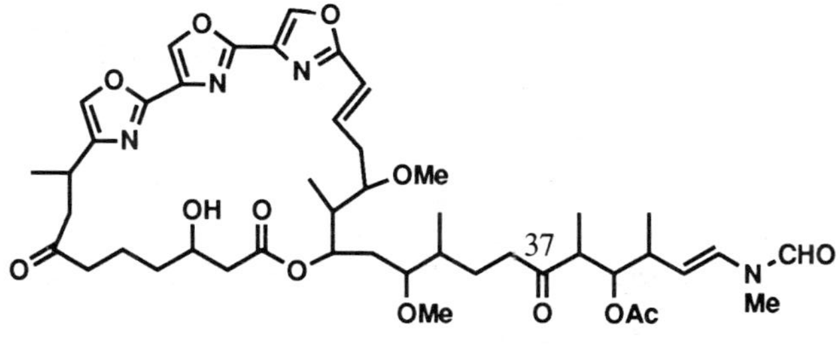

ulapualide-A (5.16)

(5.17) same as (5.16) except at C37 :

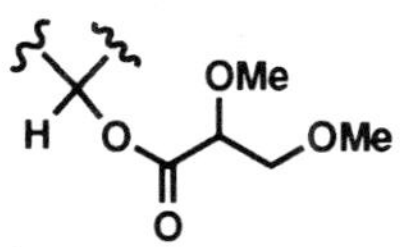

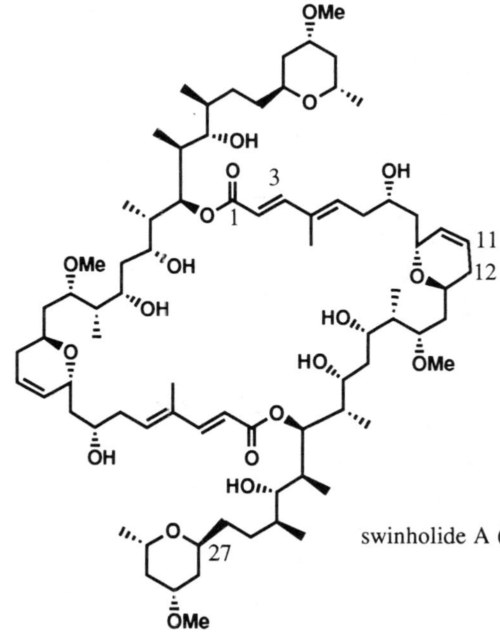

swinholide A (5.15)

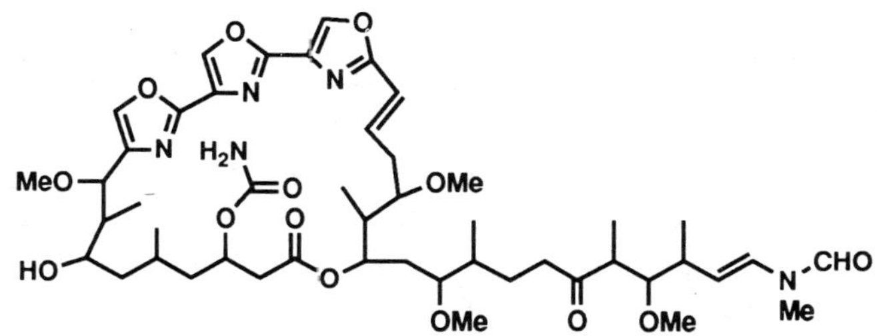

kabiramide-C (5.18)

structural similarities of the scytophycins to the higher polyketides of soil Actinomycetes, such as the macrolide antibiotics. Although no biosynthetic information exists, a visual analysis of scytophycins-A and -B suggests an origin from the combination of acetate and propionate subunits. There are clear structural and stereochemical similarities between the scytophycins (5.13 and 5.14) and the dimeric sponge metabolite swinholide-A (5.15), one of a group of homologues from this source [13]. The structure and stereochemistry along the backbones of these metabolites is identical between C-1 and C-27. Structural comparisons can also be made between the scytophycins and the ulapualides-A and -B (5.16–5.17) [14] and kabiramide-C (5.18) [15], isolated from egg masses of certain nudibranchs (mollusks). In view of these similarities it may emerge that all of these metabolites have a common cyanobacterial origin (see section 5.5).

5.2 DINOFLAGELLATES

Dinoflagellates, or marine plankton, are unicellular eukaryotic micro-organisms. They exist as unicellular species but can form large floating colonies. Many enter into symbiotic relationships with marine invertebrates such as clams, corals and sponges and are an important dietary constituent of filter-feeders. Some dinoflagellates secrete structurally complex toxins, and there are an increasing number of metabolites being isolated from higher marine organisms which probably originate from a planktonic source. Notable compounds are okadaic acid (5.19), dinophysistoxin-1 (5.20), yessotoxin (5.21) and the brevetoxins-A and -B (5.22) and (5.23). Okadaic acid (5.19) was first isolated from sponges, but subsequent investigation has established it as a true dinoflagellate metabolite [16]. Both (5.19) and (5.20) are extremely toxic and are implicated as agents responsible for diarrhetic shellfish poisoning (DSP), a gastroenteritis caused by ingestion of shellfish contaminated with marine plankton. Yessotoxin (5.21) was isolated from the digestive glands of the scallop *Patinopectin yessoensis* [17] and is also believed to be a causative agent of DSP. Perhaps the most studied dinoflagellate metabolites are the brevetoxins-A and -B (5.22) and (5.23), which have structural similarities to yessotoxin. The brevetoxins contribute to the toxicity of 'red tides' around the Gulf of Mexico and Florida. These red tides are created by blooming colonies of dinoflagellates, and the secreted toxins are an environmental hazard, killing large numbers of fish and other marine vertebrates.

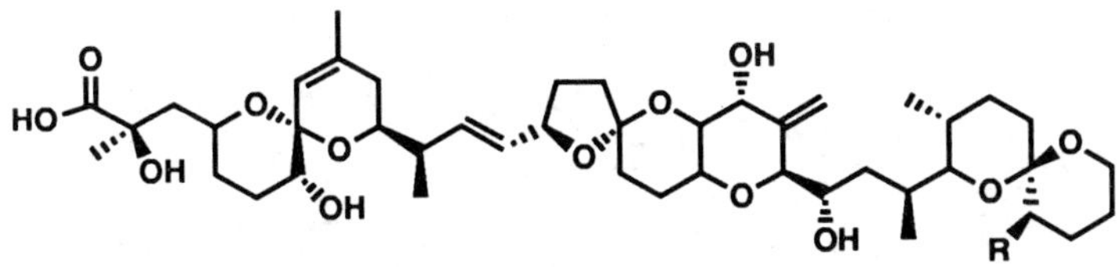

okadaic acid R = H (5.19)
dinophysistoxin-1 R = Me (5.20)

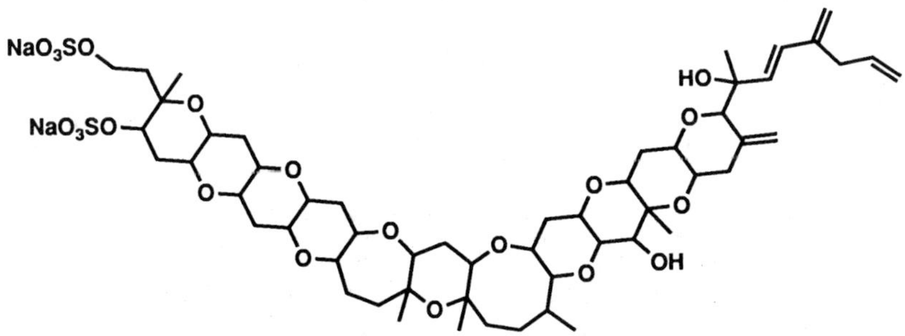

yessotoxin (5.21)

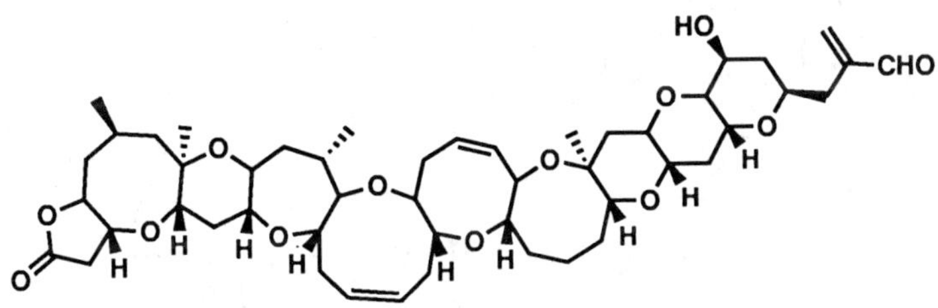

brevetoxin- A (5.22)

Norhalicondrin (5.25), has been identified from the sponge *Halichondria okadai* [18], and exhibits potent antitumour activity. In view of the structural similarity of norhalicondrin to the other complex polyethers of dinoflagellates, it would appear likely that it is actually a product of a symbiont rather than the sponge. Similarly, the pectenotoxins-1 and -2 (5.26) and (5.27) isolated from the digestive glands of the scallop *Dinophysis acuminata* [16] bear a striking structural resemblance to polyethers such as okadaic acid (5.19), and these compounds may ultimately emerge as products of dinoflagellates.

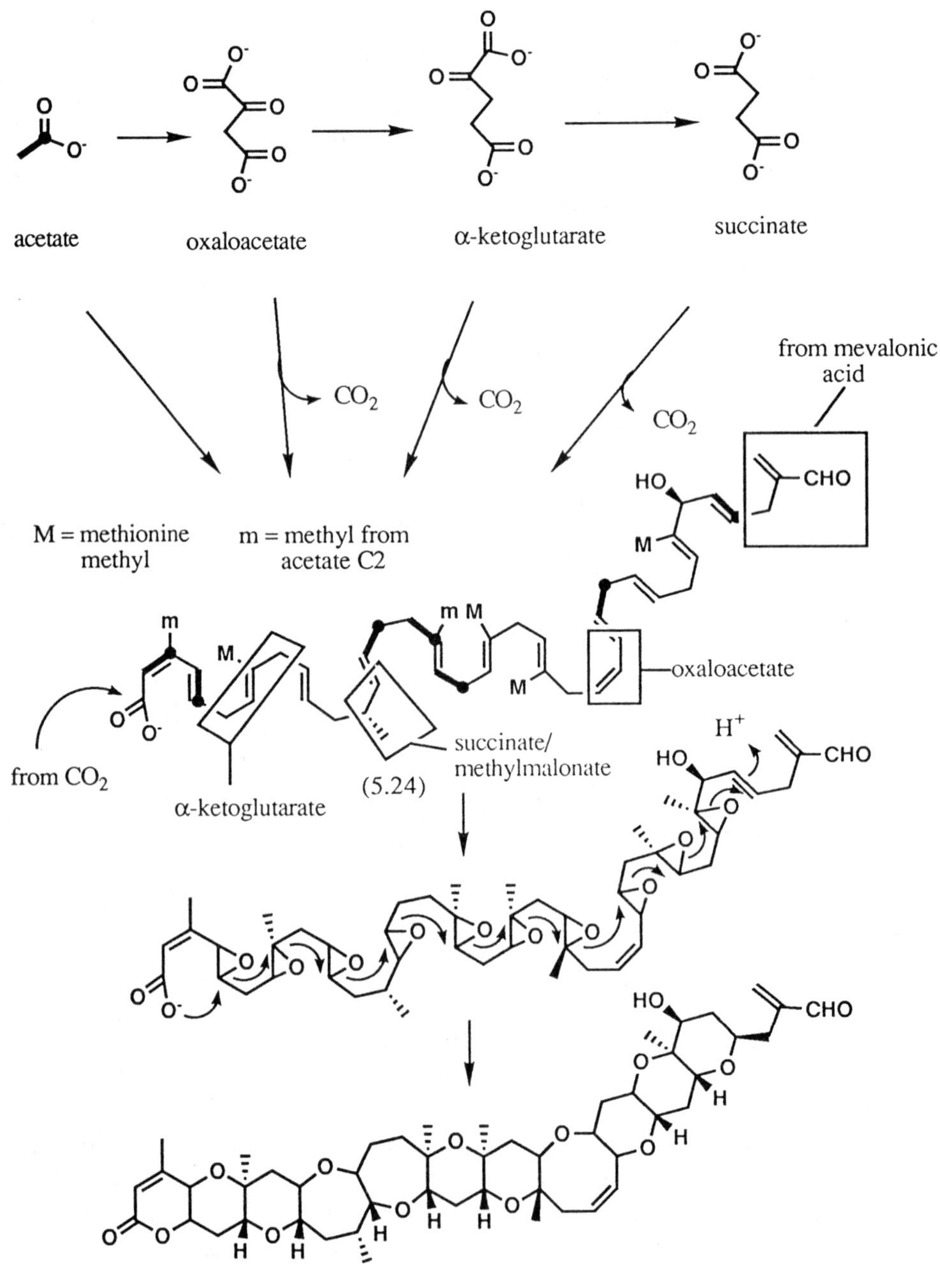

brevetoxin-B (5.23)

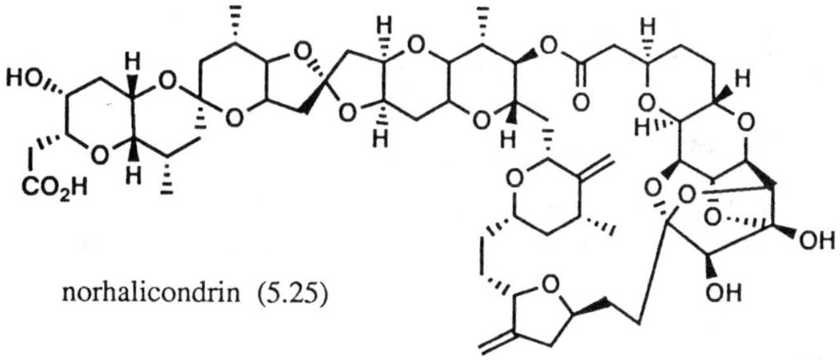

norhalicondrin (5.25)

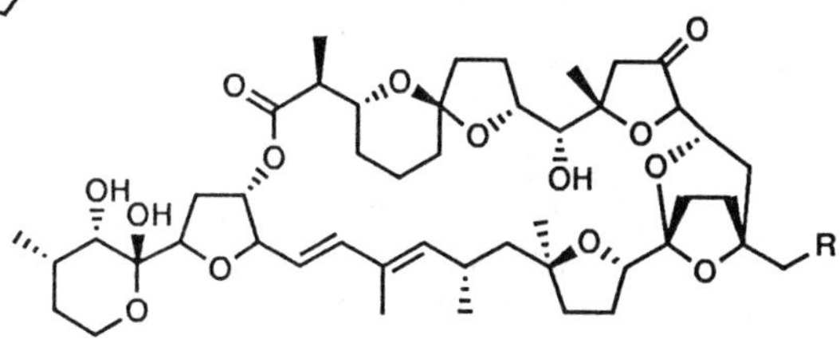

pectenotoxin 1 R = OH (5.26)
pectenotoxin 2 R = H (5.27)

Biosynthetic investigations have been carried out on the brevetoxins from *Gymnodinium breve* [19,20]. An attractive but minimum mechanism for the construction of the extensive fused ring polyether systems of the brevetoxins has been proposed by extending Westley's hypothesis for polyether biosynthesis in Actinomycetes (see section 4.4.9). For brevetoxin-B (5.23) the epoxidation of a polyene such as (5.24) followed by cascade ring closure is consistent with the location of the oxygen functionality, and it is noteworthy in this regard that 27,28-epoxybrevetoxin-B (GB-6) has been isolated from a culture of *G. breve*, suggesting epoxidase activity [21]. Labelling experiments with acetate have shown that only a limited number of intact bonds are derived from this precursor. Instead, the results are consistent with substantial involvement of intermediates on the citric acid cycle such as oxaloacetate, α-ketoglutarate and succinate. These species, or their thioesters, could be involved with malonate and undergo a series of decarboxylative condensations to generate the backbone of the brevetoxins. In the case of brevetoxin-B (5.23), seven intact acetate units are involved. The decarboxylation of α-ketoglutarate gives rise to a C_4 subunit (boxed in (5.24)), and four of the terminal carbons are believed to arise from mevalonic acid. Methionine is also involved, donating, in the case of brevetoxin-B (5.23), four of the seven branched methyl groups. Two of the methyl groups arise from C-2 of acetate and are probably left isolated after decarboxylation of the carboxylate group of acetate. Carbon-1 is believed to arise from CO_2, and the remaining carbons of brevetoxin-B originate from C_3 subunits derived from the decarboxylation of oxaloacetate to give linear C_3 units and the conversion of succinyl-CoA to methylmalonyl-CoA to generate a 'propionate' primer and incorporation of a branching C_3 unit. An example is highlighted for each case in polyene (5.24).

The amphidinolides (5.28–5.31) are an interesting group of macrocyclic lactones which have been isolated from cultures of *Amphidinium* sp. [22]. Amphidinolides A–C were identified from a single dinoflagellate isolated from the gut of an Okinawan (Japan) flat worm, whereas amphidinolide-E (5.31) is the product of a separate *Amphidinium* sp. from the same source. They are structurally analogous metabolites, although not obvious homologues of each other, and the origin of the variations in their carbon skeleton must await appropriate biosynthetic investigations. The branching methyl and exo-methylene functionality in each of the metabolites is random and inconsistent with an origin from propionate. Several possibilities exist, including acetate/methionine combinations however it may emerge that the amphidinolides have a closer relationship to the brevetoxins, where skeletal subunits are derived from acetate/methionine and citric acid cycle intermediates.

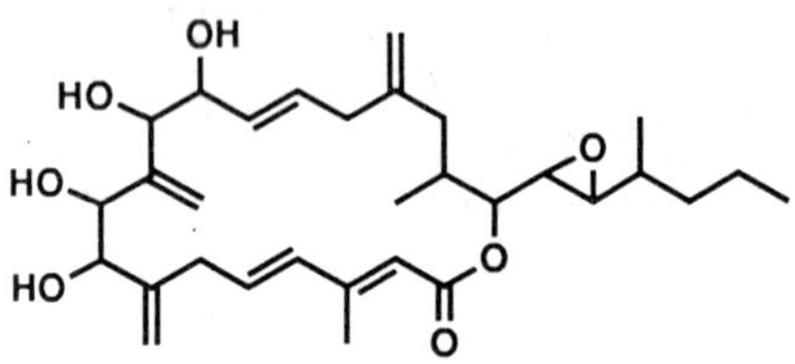

amphidinolide A (5.28)

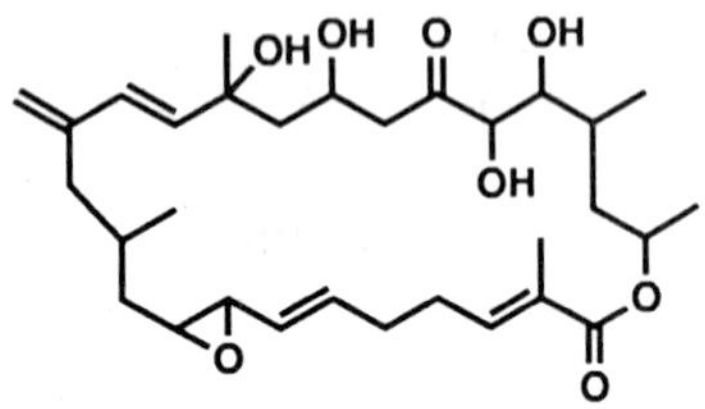

amphidinolide B (5.29)

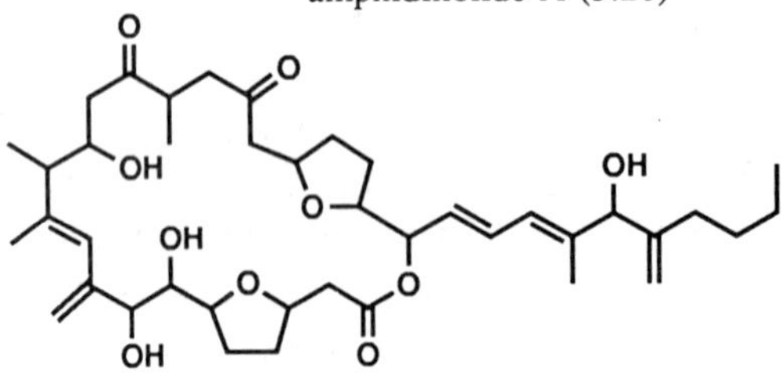

amphidinolide C (5.30)

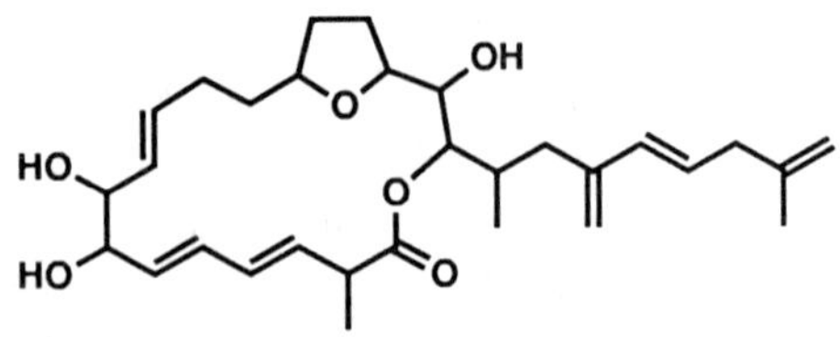

amphidinolide E (5.31)

5.3 ALGAE

The algae are marine plants. They elaborate a great variety of terpene-based compounds, but like terrestrial plants are not particularly versatile polyketide producers. They do, however, generate a bewildering array of complex lipids, most of which are believed to arise from the modification of unsaturated fatty acids. For example, compounds such as the alkylated resorcinol (5.32) and the acylated phloroglucinol (5.33) have been identified from several species of brown algae (seaweeds) [1a] and these are structurally analogous to other simple aromatic polyketides from plants (see section 3.1).

A variety of C_{11} acyclic and cyclic hydrocarbons act as sex pheromones during the reproduction of brown algae. Dictyopterene-B (5.34) is one of the most ubiquitous examples of this class of volatile hydrocarbons. Viridiene (5.35) and ectocarpene (5.36) from a *Syringoderma* sp. and *Ectocarpus siliculosus* respectively act as male gamete releasing/attracting compounds. Interestingly, ectocarpene (5.36) is also a component of the essential oils of the flowering plant *Senecio isatideus* from South Africa. Biosynthetic investigations of the algae are impractical, but a study of the plant has revealed the intermediacy of dodeca-3,6,9-trienoic acid (5.37) during ectocarpene biosynthesis and a pathway has been proposed [23]. The origin of viridiene (5.35) can be rationalized via an alternative cyclization mode. The occurrence of ectocarpene from each system clearly points to a similar, if not identical, mode of biosynthesis.

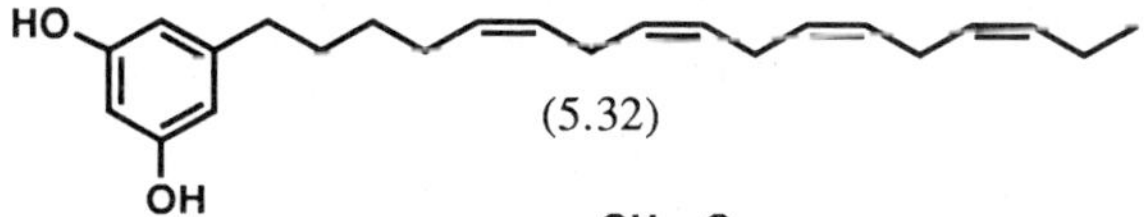

(5.32)

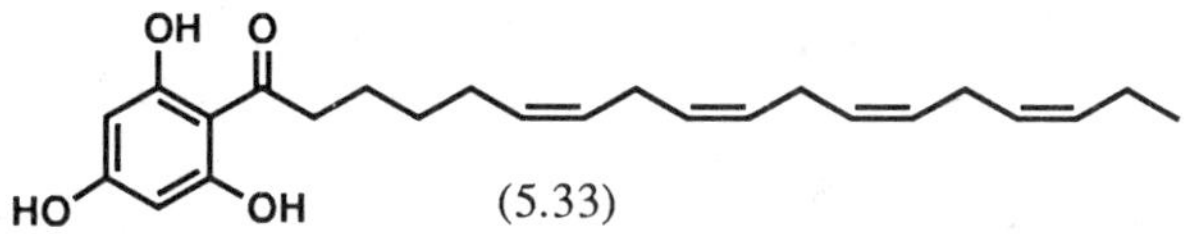

(5.33)

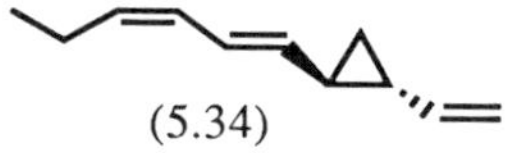

(5.34)

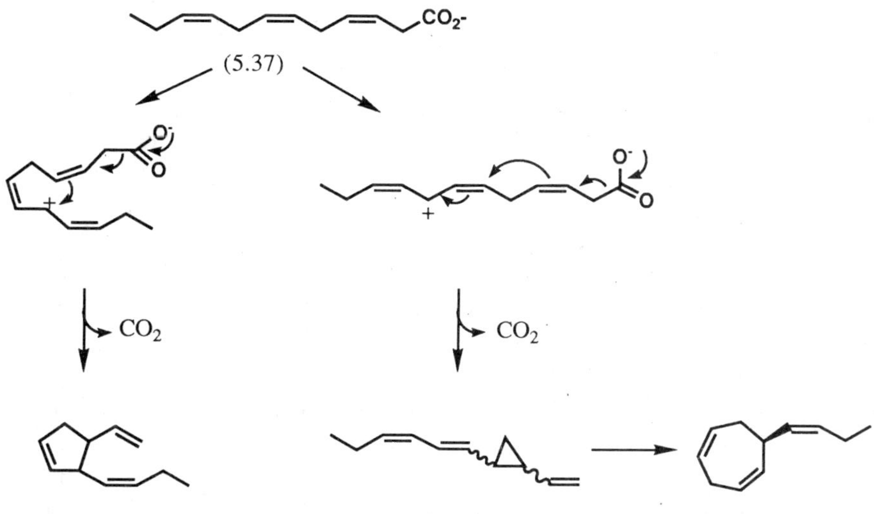

viridiene (5.35)

ectocarpene (5.36)

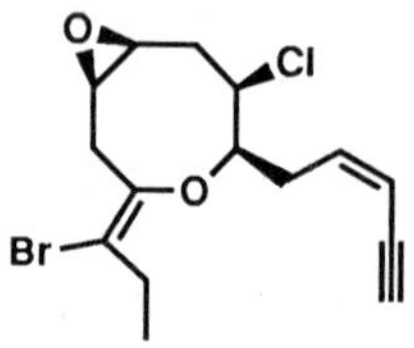

cis-epoxyrhodophytin (5.38)

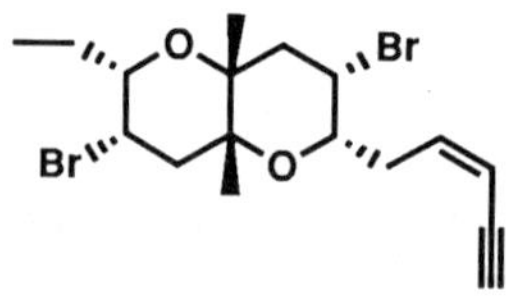

elatenyne (5.39)

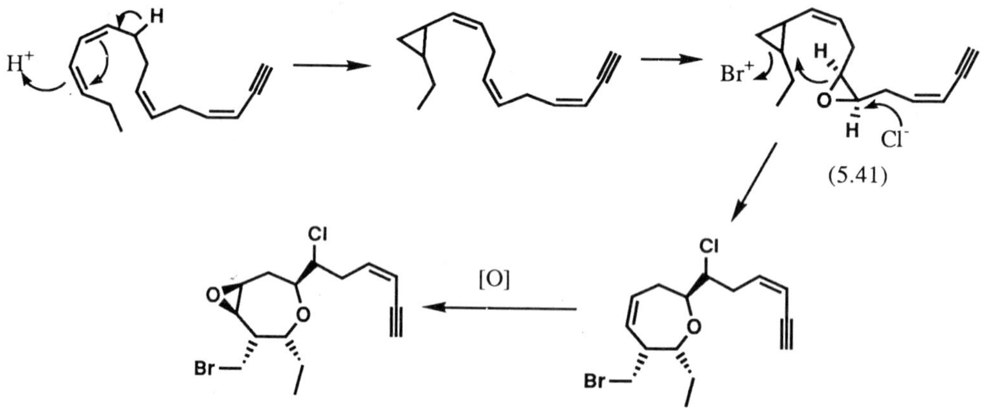

(5.41)

rogiolenyne-A (5.40)

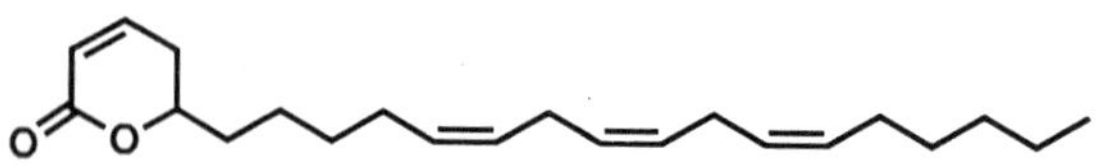

(5.42)

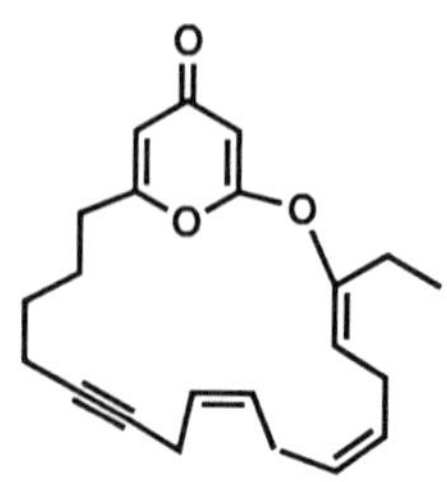

(5.43)

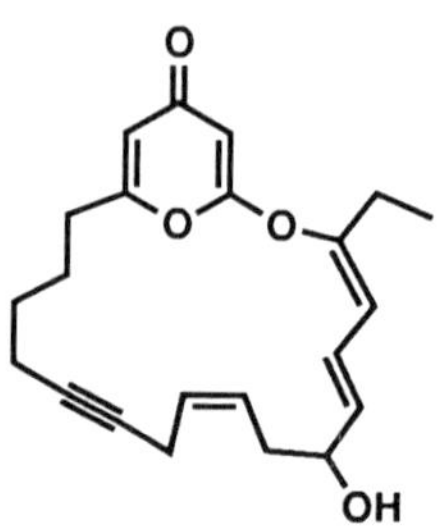

(5.44)

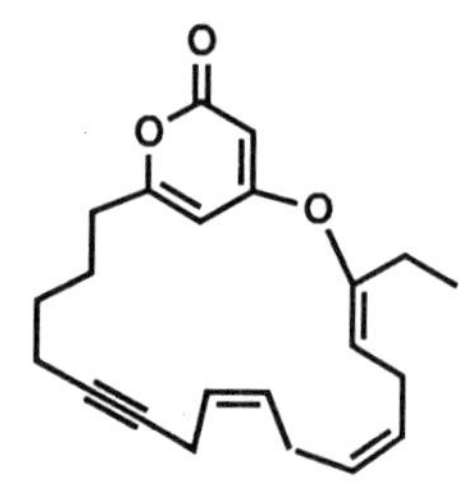

(5.45)

The red algae of *Laurencia* secrete numerous C_{15} halogenated lipids such as *cis*-epoxyrhodophytin (5.38), elatenyne (5.39) and rogiolenyne A (5.40). Elatenyne (5.39) possesses a bis-fused pyran ring system, common in the dinoflagellate metabolites, which probably arises via cyclization of an intermediate epoxide. Rogiolenyne-A (5.40) has an unusual branched C_{15} skeleton, and a biosynthetic rationale has been proposed involving a cyclopropane–epoxide intermediate (5.41), although this mechanism has no experimental foundation [24].

The C_{22} lactone (5.42) and the macrocyclic γ-pyrones (5.43)–(5.45) have all been identified from *Phacelocarpus labillardieri*, one of the most common red algae in Southern Australian waters [25]. The lactone (5.42) may, or may not, be an intermediate in the biosynthesis of the γ-pyrones, but its co-occurrence strongly supports their origin from similarly modified C_{20} and C_{22} fatty acids.

5.4 MOLLUSKS

Mollusks encompass a variety of marine animals, the most common being the bivalves (clams, mussels, scallops, oysters, etc.) and the gastropods such as the snails and related creatures. In terms of polyketide metabolites the most interesting mollusks are the air-breathing gastropods of the genus *Siphonaria,* which inhabit the intertidal zone. These organisms secrete an ever-increasing variety of metabolites derived exclusively from propionate. For example, the origin of muamvatin (5.46), the denticulatins-A and -B (5.47) and (5.48) and the siphonarins-A and -B (5.49) and (5.50) can all be rationalized in terms of a polypropionate skeleton. In support of this, a biosynthetic investigation has demonstrated that [1-^{14}C]-propionate is incorporated into the denticulatins-A and -B from *Siphonaria denticulata* [26]. In this study the denticulatins were extracted from the foot mantle of the gastropod and were not evident from extracts of the digestive organs, implying that they were not accumulated through dietary intake.

An analysis of the structure of these compounds reveals that there are common structural and stereochemical homologies between denticulatin-A, the siphonarins and muamvatin when they are drawn linearly in their open chain forms [27]. These correlations are only valid if the polyketide chains are assembled in the same direction, and this is difficult to establish as the terminal carboxylate group is absent in each case. The correlations are interesting, however, and do suggest that similar polyketide synthases are operating and that they are biogenetically related to each other.

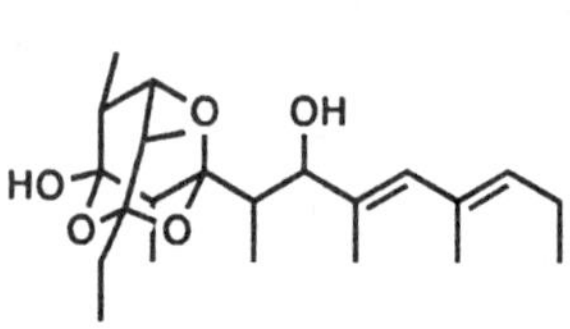

muamvatin (5.46)

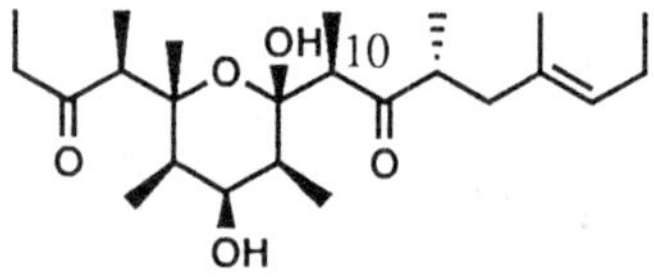

denticulatin A (5.47)
denticulatin B epimeric at C10 (5.48)

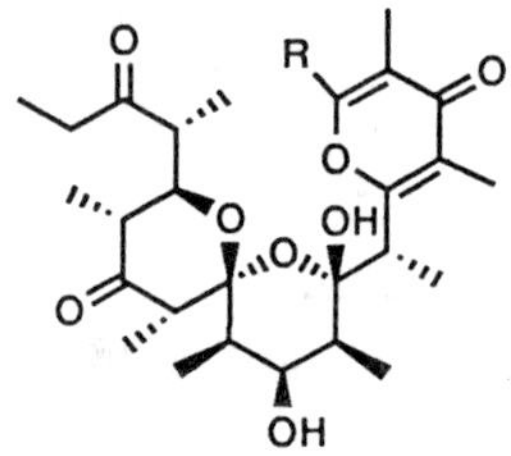

R = Me (5.49)
R = Et (5.50)

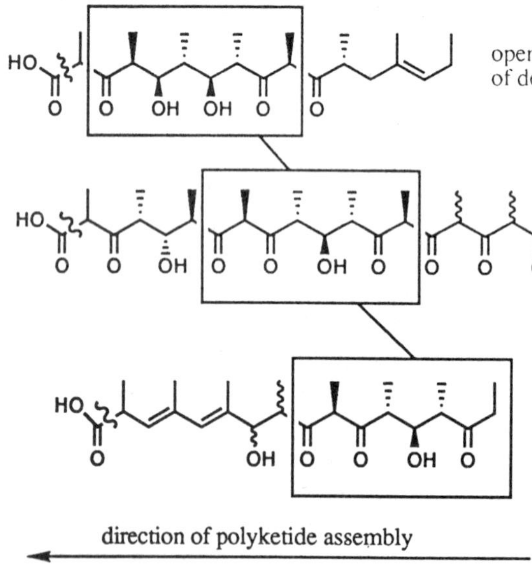

open chain form of denticulatin A (5.47)

open chain form of the siphonarins (5.49) and (5.50)

open chain form of muamvatin (5.46)

The baconipyrones-A to -D (5.51–5.54) are an interesting group of polypropionate metabolites which do not possess a contiguous carbon skeleton and are believed to be derived from the siphonarins [28]. The co-occurrence of siphonarin-A (5.49) with the bacinopyrones in *Siphonaria baconi* led to the proposal that a retro-aldol condensation of the siphonarin skeleton could account for baconipyrones-C and -D and that a subsequent intramolecular aldol condensation would give rise to baconipyrones-A and -B.

Other polypropionates from *Siphonaria* spp. are pectinatone (5.55), dimenensin-A (5.56) and siphonarienefuranone (5.57). Some of these compounds display anti-bacterial activity, and pectinatone (5.5) and siphonarienefuranone (5.57) also inhibit the growth of yeast [29]. The absolute stereochemistry of the branching methyl groups is uniform for this siphonarid series, each metabolite having the *S*-configuration at the asymmetric centres along the backbone [27,29]. The polypropionate metabolites, aglajnes 1–3

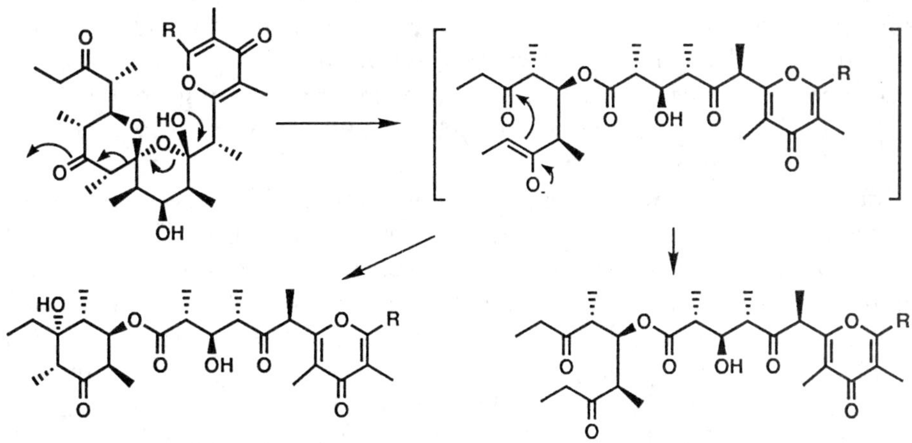

baconipyrone A R = Et (5.51)
baconipyrone B R = Me (5.52)

baconipyrone C R = Et (5.53)
baconipyrone D R = Me (5.54)

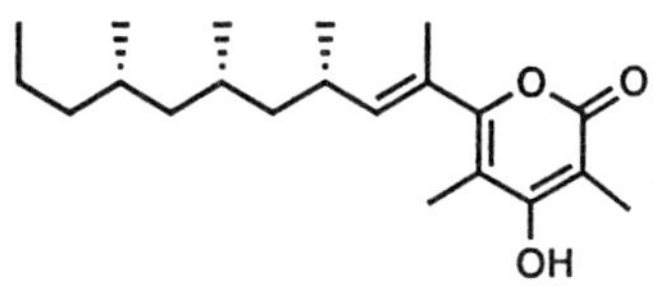

pectinatone (5.55)

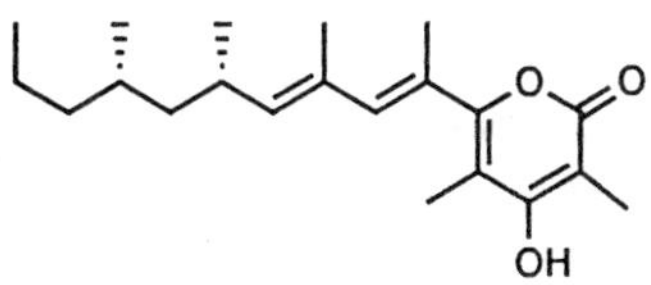

dimenensin A (5.56)

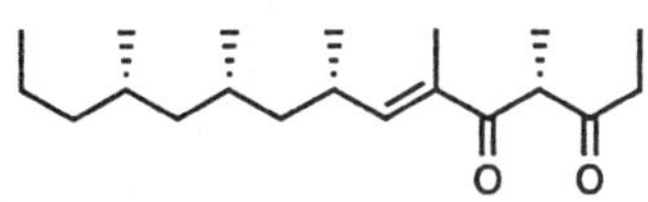

siphonarienefuranone (5.57)

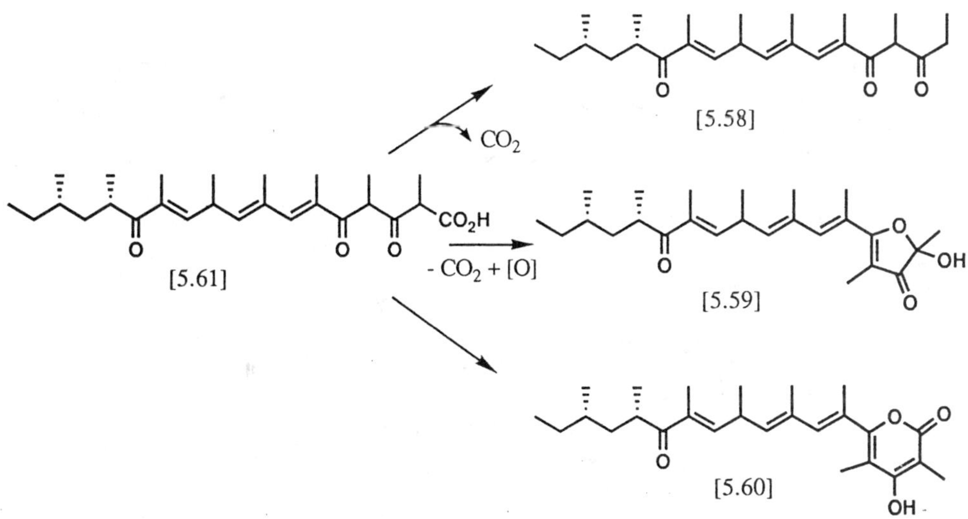

(5.58–5.60), from a different class of mollusk, *Aglaja depicta,* have the same absolute configuration at the stereochemically defined centres [30], consistent with a similar polyketide synthase operating in each case. The co-occurrence of aglajnes 1–3 can be rationalized in terms of a single polypropionate precursor, such as (5.61), which cyclizes in various ways to generate the individual metabolites.

The γ-pyrone functionality is a common feature of mollusk polypropionates, and other examples are the vallartanones-A and -B (5.62) and (5.63) from *Siphonaria maura* [31]. More complex carbon skeletons are represented by tridachione (5.64), 9,10–deoxytridachione (5.65), crispatone (5.66) and crispatene (5.67), metabolites of the Sacoglossan *Tridachia crispata* [32]. The Sacoglossans originate from a different gastropod subclass to the Siphonarins but these metabolites can also be interpreted in terms of a polypropionate origin. It has been demonstrated *in vitro* that 9,10-deoxytridachione (5.65) undergoes a facile rearrangement to photodeoxytridachione (5.68) in sunlight, and therefore the bis-fused ring system of crispatone (5.66) and crispatene (5.67) is believed to arise photochemically [32].

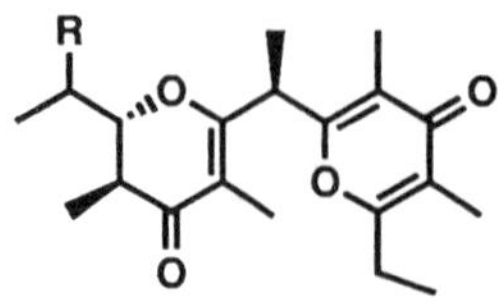

R = Me (5.62)
R = H (5.63)

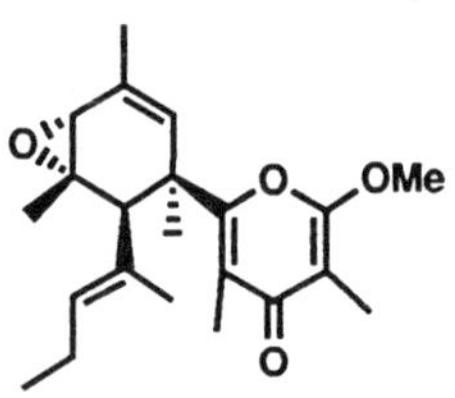

(5.64)

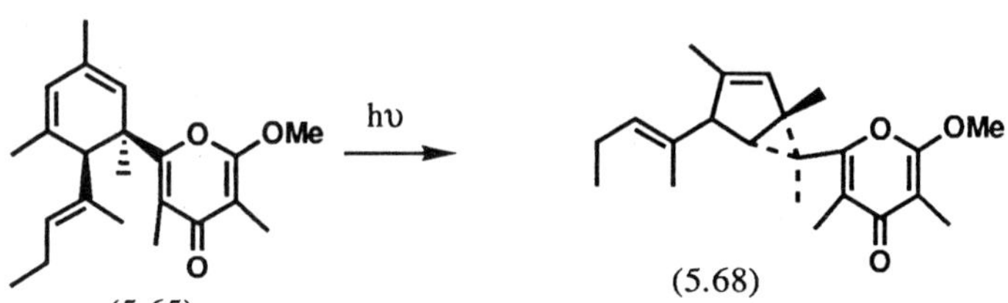

(5.65) (5.68)

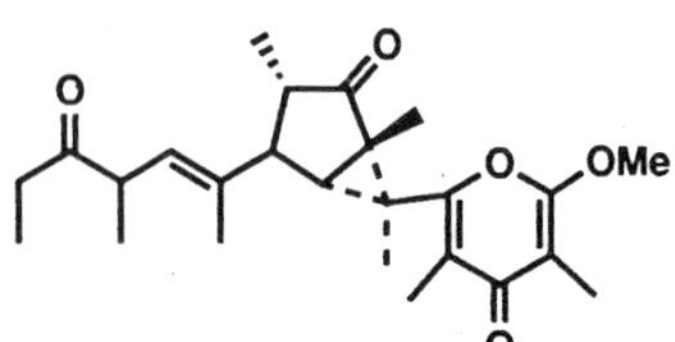

(5.66)

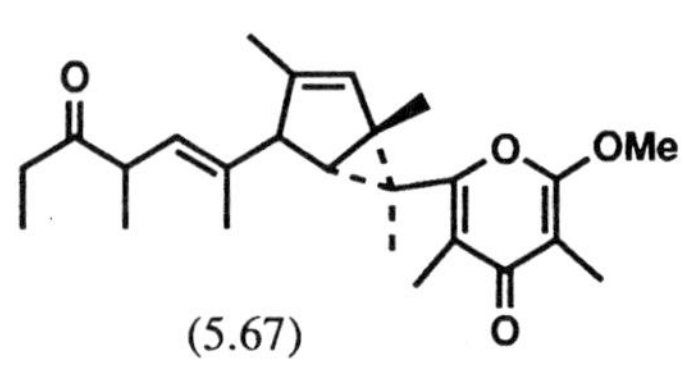

(5.67)

It is interesting that the only other polypropionate polyketides are the erythromycin antibiotics of terrestrial prokaryotes (Actinomycetes), and presumably these marine metabolites are biosynthesized in a similar manner involving methylmalonyl-CoA or a related thioester (see section 4.4.2). Whether the marine polypropionates are products of symbiotic marine bacteria remains an open and intriguing question. A formal endosymbiosis in line with Margulis's theory of cellular evolution [33], where prokaryotes are precursors to eukaryotic organelles, may implicate the dinoflagellates as originators of the polypropionate metabolites. This view is questionable, however, as the known dinoflagellate polyketides, such as the brevetoxins and the amphidinolides, are structurally more elaborate, and the limited biosynthetic evidence suggests a modified polyketide pathway (see section 5.2). Whether mollusk, prokaryote or dinoflagellate, the answer to these questions will have to await appropriate microbiological analysis to determine the nature and examine the role of symbiotic species present in mollusk tissue.

5.5 SPONGE OR MICRO-ORGANISM?

It is very difficult to rigorously define the true origin of marine polyketides, or metabolites in general, from a sponge. This is a consequence of the filter-feeding habits and symbiotic relationships characteristic of sponges. On the other hand, structural comparisons of metabolites isolated from sponges with metabolites of known origin can imply a dietary or symbiotic source. Norhalichondrin (5.25) has already been discussed in this regard as a probable dinoflagellate metabolite (see section 5.2). The obvious structural and stereochemical similarities between swinholide-A (5.15), isolated from the sponge *Theonella swinhoei* [13], and the scytophycins (5.13) and (5.14) from blue-green algae is persuasive evidence that these metabolites have a common blue-green algal origin. *Theonella swinhoei* is known to have symbiotic blue-green algae associated with it, and this would appear to be the most likely source of swinholide-A. A similar argument can be made for the ulapualides (5.16) and (5.17) and kabiramide-C (5.18) from nudibranch eggmasses (see section 5.1).

The geodiamolides-A and -B (5.69) and (5.70) [34] from a Caribbean species of *Geodia* and jaspamide (5.71) from *Jaspis sp.* collected in Fiji [35] are structurally related sponge metabolites which appear to have a mixed peptide/polypropionate biogenesis. Their polypropionate moieties are identical in each case. The origin of these metabolites from the sponges has not been shown rigorously and the presence of the polypropionate moiety tends to suggest that this fragment at least has a similar origin to those polypropionates found in mollusks.

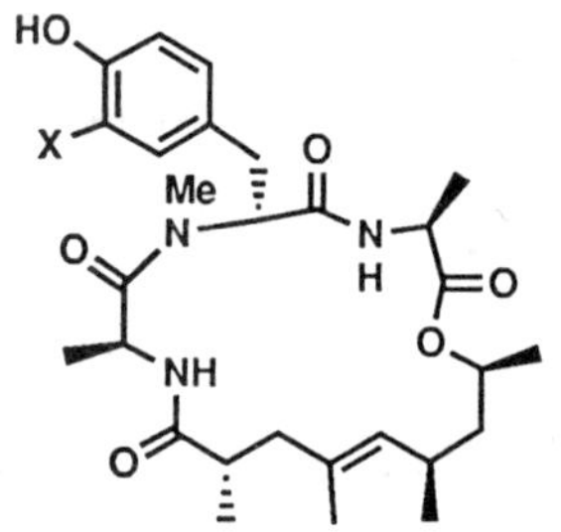

R = I (5.69)
R = Br (5.70)

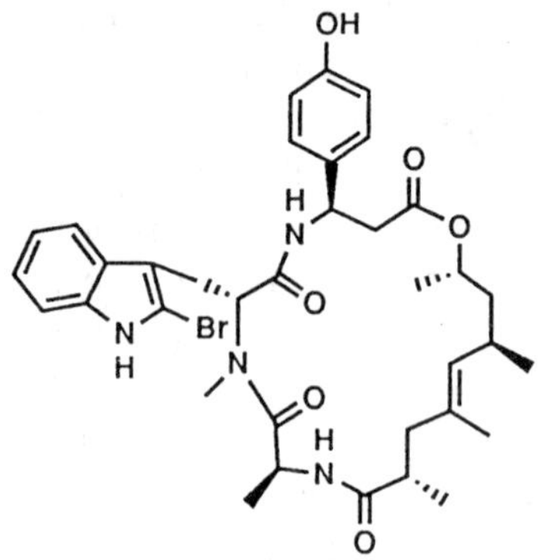

(5.71)

Tedanolide (5.72) is a macrocyclic lactone which has been isolated in the Caribbean from *Tedania ignis,* a sponge known locally as 'fire sponge' owing to the skin irritation it commonly inflicts after handling [36]. The structure of tedanolide is similar to the polyether and macrolide antibiotics of terrestrial bacteria, and the origin of its framework can be rationalized from acetate and propionate subunits. Intriguingly, if the structure and stereochemisty along the backbone of tedanolide is compared with that of the polyether antibiotics of Actinomycetes, similarities are apparent [37]. These are most obvious if the putative diepoxide precursor believed to be involved in the biosynthesis of the bisdispiroketal polyether noboritomycin (see section 4.5) is compared with tedanolide (5.72). There exists an eight-subunit sequence which differs only in one instance (acetate for propionate), and of the eight matching asymmetric centres there are only two inconsistencies one of which occurs at a readily epimerizable centre in tedanolide [38]. Focusing on the functionality, there are three matching ketones and an epoxide with identical regio- and stereochemistry. It is difficult to rationalize these similarities as coincidental and the origin of tedanolide as a true sponge metabolite is brought into question. Tedanolide is not typical of the dinoflagellate metabolites and structurally (and probably biosynthetically) appears closer to the higher polyketides of soil Actinomycetes or even the polypropionates of the mollusks. Again, the existence of symbiotic marine bacteria, possibly with a close taxonomic relationship to soil bacteria, emerges as a possible scenario for the origin of tedanolide.

The highly coloured sponge *Latrunculia magnifica* proliferates in the Red Sea and appears to be immune to predators. If squeezed it exudes a red fluid containing neurotoxin activity which repels fish, or kills them in minutes when the experiment is carried out in an aquarium. The latrunculins A–D (5.73–5.76) have been isolated and are toxic constituents of this sponge [39].

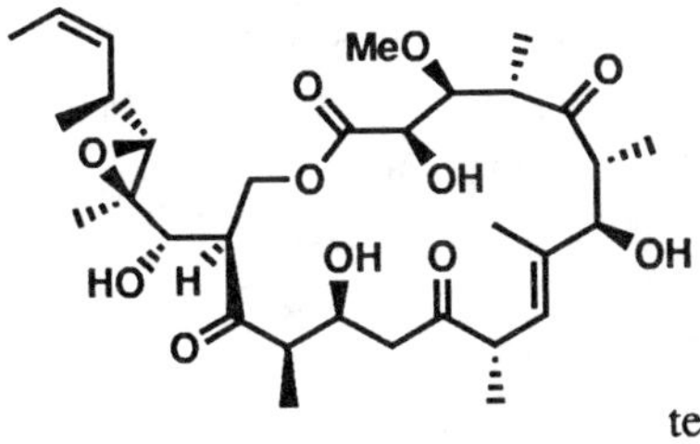

tedanolide (5.72)

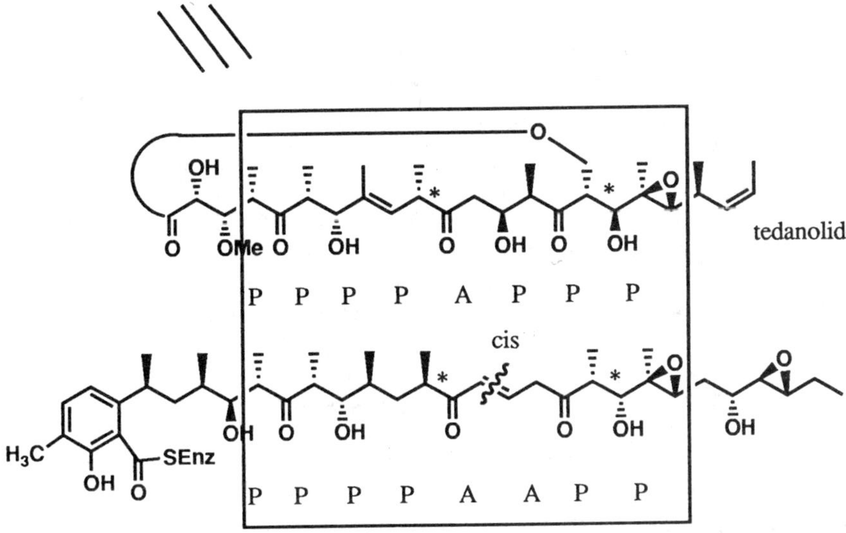

tedanolide (5.72)

diepoxide precursor to noboritomycin-A

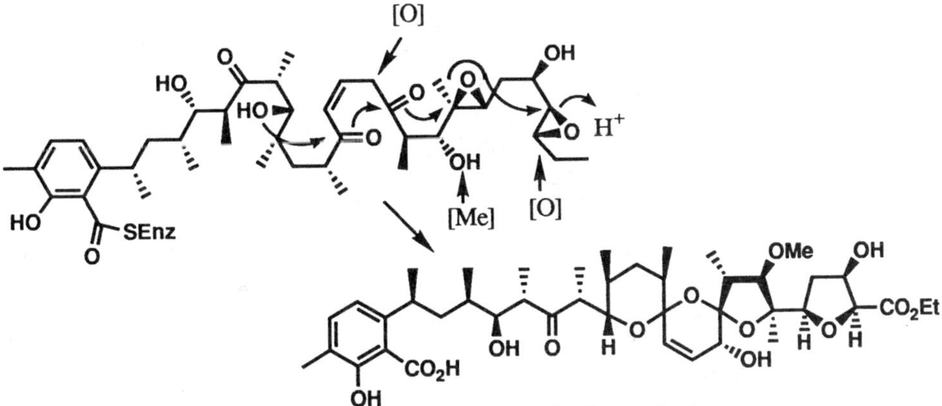

noboritomycin-A

These compounds possess a thiazolone ring system which probably derives from cysteine, and the biosynthesis of the latrunculins can be rationalized in terms of a polyketide chain, a cysteine residue and a C_1 unit contributing the carbonyl functionality of the thiazolone ring. A functionalized thiazole, mycothiazole (5.77), has been isolated from the Fijian sponge *Spongia mycofijiensis,* and a polyketide/cysteine biogenetic relationship has been proposed linking mycothiazole with the latrunculins [40].

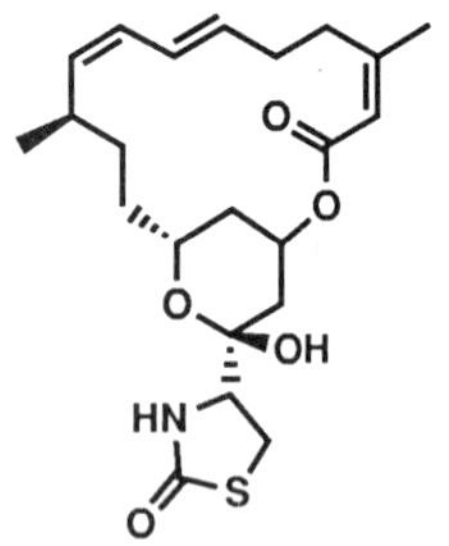

latrunculin A (5.73)

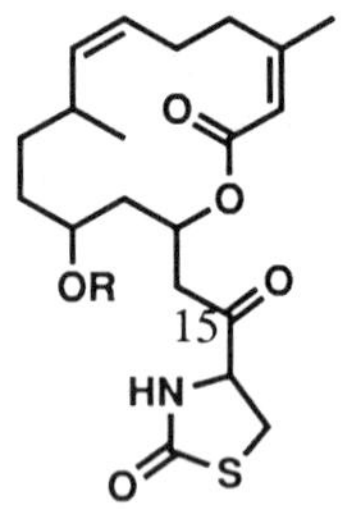

latrunculin C R = Me (5.75)
latrunculin D R = H and

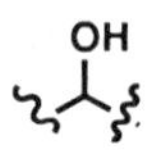

at C 15 (5.76)

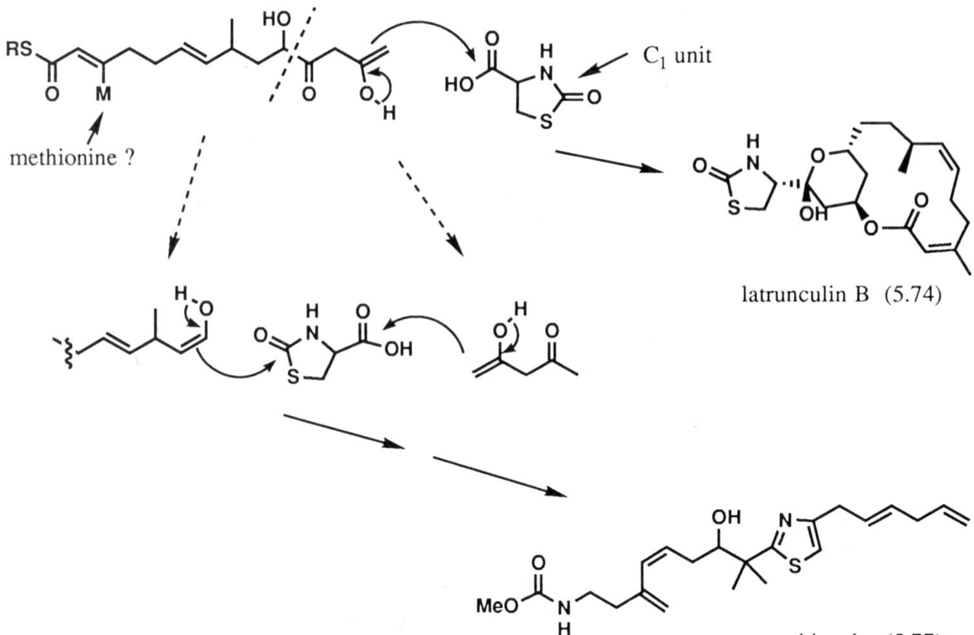

latrunculin B (5.74)

mycothiazole (5.77)

REFERENCES

[1] For excellent reviews on marine metabolite structures see (a) D. J. Faulkner, *Nat. Prod. Rep.*, 1984, **1**, 251; *ibid.*, p. 551; (b) D. J. Faulkner, *Nat. Prod. Rep.*, 1986, **3**, 1; (c) D. J. Faulkner, *Nat. Prod. Rep.*, 1987, **4**, 539; (d) D. J. Faulkner, *Nat. Prod. Rep.*, **5**, 613.

[2] M. J. Garson, Biosynthetic studies on marine natural products. *Nat. Prod. Rep.*, 1989, **6**, 143.

[3] J. J. Lee, P. M. Dewick, C. P. Gorst-Allman, F. Spreafico, C. Kowal, C.-J. Chang, A. McInnes, J. A. Walter, P. J. Keller and H. G. Floss, Further studies on the biosynthesis of the boron containing antibiotic aplasmomycin. *J. Am. Chem. Soc.*, 1987, **109**, 5426.

[4] G. S. Holland, D. D. Jamieson, J. L. Reichelt, G. Viset and R. J. Wells, Three aromatic acids from a marine bacterium. *Chem. Ind.* (London), 1984, **23**, 850.

[5] T. Kitahara, H. Naganawa, T. Okazaki, Y. Okami and H. Umezawa, Structure of SS-228Y, an antibiotic from *Chainia* species. *J. Antibiotics*, 1975, **28**, 280.

[6] Y. Tamura, F. Fukata, M. Sasho, T. Tsugoshi and Y. Kita, Synthesis of antibiotic SS-228R. Strong base induced cycloaddition of homophthalic anhydrides. *J. Org. Chem.*, 1985, **50**, 2273.

[7] N. Imamura, K. Katinuma, N. Ikekawa, H. Tanaka and S. Omura, Biosynthesis of vineomycins A_1 and B_2. *J. Antibiotics*, 1982, **35**, 602.

[8] J. H. Cardellina II and R. E. Moore, Malyngic acid, a new fatty acid from *Lyngbya majuscula*. *Tetrahedron*, 1980, **36**, 993.

[9] J. H. Cardellina II, D. Dalietos, F.-J. Marner, J. S. Mynderse and R. E. Moore, (−)-trans-7-(*S*)-methoxytetradec-4-enoic acid and related amides from the marine cyanophyte *Lyngbya majuscula*. *Phytochemistry*, 1978, **17**, 2091.

[10] G. R. Pettit, Y. Kamano, C. L. Herald, C. Dufresne, R. B. Bates, J. M. Schmidt, R. L. Cerny and H. Kizv, Antineoplastic agents 190. Isolation and structure of the cyclodepsipeptide dolstatin-14. *J. Org. Chem.*, 1990, **55**, 2989.

[11] J. H. Cardellina II, R. E. Moore, E. V. Arnold and J. Clardy, Structure and absolute configuration of malyngolide, an antibiotic from the marine blue-green alga *Lyngbya majuscula* Gomont. *J. Org. Chem.*, 1979, **44**, 4039.

[12] M. Ishibashi, R. E. Moore, G. M. L. Patterson, C. Xu and J. Clardy, Scytophycins, cytotoxic antimycotic agents from the cyanophyte *Scytonema pseudohofmanni*. *J. Org. Chem.*, 1986, **51**, 5300.

[13] M. Doi, T. Ishida, M. Kobayashi and I. Kitagawa, Molecular conformation of swinholide A, a potent cytotoxic dimeric macrolide from the Okinawan marine sponge *Theonella swinhoei*: X-ray crystal structure of

its diketone derivative. *J. Org. Chem.*, 1991, **56**, 3629.

[14] J. A. Roesener and P. J. Scheuer, Ulapualide A and B, extraordinary antitumor macrolides from nudibranch eggmasses. *J. Am. Chem. Soc.*, 1986, **108**, 846.

[15] S. Matsunaga, N. Fusetani, K. Hashimoto, K. Koseki and M. Noma, Kabiramide C, a novel antifungal macrolide from nudibranch eggmasses. *J. Am. Chem. Soc.*, 1986, **108**, 847.

[16] T. Yasumoto, M. Murata, Y. Oshima, M. Sano, G. K. Matsumoto and J. Clardy, Diarrhetic shellfish toxins. *Tetrahedron,* 1985, **41**, 1019.

[17] M. Murata, M. Kumagai, J. S. Lee and T. Yasumoto, Isolation and structure of yessotoxin, a novel polyether compound implicated in diarrhetic shell fish poisoning. *Tetrahedron Letts.*, 1987, **47**, 5869.

[18] D. Uemura, K. Takahashi, T. Yamamoto, C. Katayama, J. Tanaka, Y. Okumura and Y. Hirata, Norhalichondrin A: an antitumour polyether macrolide from a marine sponge. *J. Am. Chem. Soc.*, 1985, **107**, 4796.

[19] H.-N. Chou and Y. Shimizu, Biosynthesis of brevetoxins. Evidence for the mixed origin of the backbone carbon chain and the possible involvement of dicarboxylic acids. *J. Am. Chem. Soc.*, 1987, **109**, 2184.

[20] M. S. Lee, G.-W. Qui, K. Nakanishi and M. G. Zargoski, Biosynthesis of the brevetoxins, potent neurotoxins produced by the dinoflagellate *Gymnodinium breve. J. Am. Chem. Soc.*, 1989, **111**, 6234.

[21] H. N. Chou, Y. Shimizu, G. Van Duyne and J. Clardy, Isolation and structures of two new polycyclic ethers from *Gymnodinium breve* Davis (=*Ptychodiscus brevis*). *Tetrahedron Letts.*, 1985, **26**, 2865.

[22] J. Kobayashi, M. Ishibashi, T. Murayama, M. Takamatsu, M. Iwamura, Y. Ohizumi and T. Sasaki, Amphidinolide E. A novel antileukemic 19-membered macrolide from the cultured symbiotic dinoflagellate *Amphidinium* sp. *J. Org. Chem.*, 1990, **55**, 3421.

[23] W. Boland and K. Mertes, Biosynthesis of algal pheromones. A model study with the composite. *Senecio isatideus. Eur. J. Biochem.*, 1985, **147**, 83.

[24] G. Guella and F. Petra, Rogiolenyne A, B and C: the first branched marine C_{15} acetogenins. Isolation from the red seaweed *Laurencia microcladia* or the sponge *Spongis zimocca* of Il Rogiolo. *Helv. Chim. Acta.*, 1991, **74**, 47.

[25] R. Kazlauskas, P. T. Murphy, R. J. Wells, and A. J. Blackman, Macrocyclic enol-ethers containing an acetylenic group from the red alga *Phacelocarpus labillardieri, Aust. J. Chem.*, 1982, **35**, 113.

[26] D. C. Manker, M. J. Garson and D. J. Faulkner, *De novo* biosynthesis of polypropionate metabolites in the marine pulmonate *Siphonaria denticulata. J. Chem. Soc. Chem. Commun.*, 1988, 1061.

[27] M. J. Garson, C. J. Small, B. W. Skelton, P. Thinapong and A. H.

White, Structural and stereochemical correlations of the polypropionate metabolites from the marine pulmonates: revision of the relative stereochemistry of pectinatone by X-ray structure analysis. *J. Chem. Soc. Chem. Commun.*, 1990, 805.

[28] D. C. Manker, D. J. Faulkner, T. J. Stout and J. Clardy, The baconipyrones: novel polypropionates from the pulmonate *Siphonaria baconi*. *J. Org. Chem.*, 1989, **54**, 5371.

[29] M. Norte, F. Cataldo, A. G. Gonzalez, M. L. Rodriguez and C. Ruiz-Perez, New metabolites from the marine mollusc *Siphonaria grisea*. *Tetrahedron*, 1990, **46**, 1669.

[30] G. Cimino, G. Sodano and A. Spinella, New propionate derived metabolites from *Aglaja depicta* and from its prey *Bulla striata* (Opisthobranch Molluscs). *J. Org. Chem.*, 1987, **52**, 5326.

[31] D. C. Manker and D. J. Faulkner, Vallartanones-A and -B, polypropionate metabolites of *Siphonaria maura* from Mexico. *J. Org. Chem.*, 1989, **54**, 5374.

[32] C. Ireland and P. J. Scheuer, Photosynthetic marine mollusks: *in vivo* ^{14}C incorporation into metabolites of the Sacoglossan *Placobranchus ocellatus*. *Science*, 1979, **205**, 922.

[33] L. Margulis, *Symbiosis in cell evolution*. W. H. Freeman, 1981, San Francisco.

[34] W. R. Chan, W. F. Tinto, P. S. Manchand and L. J. Torado, Stereostructures of geodiamolides A and B, novel cyclodepsipeptides from the marine sponge *Geodia* sp. *J.Org. Chem.*, 1987, **52**, 3091.

[35] T. M. Zabriskic, A. Klocke, C. M. Ireland, A. H. Marcus, T. F. Molinski, D. J. Faulkner, C. Xu and J. C. Clardy, Jaspamide, a modified peptide from the *Jaspis* sponge with insectecidal and antifungal activity. *J. Am. Chem. Soc.*, 1986, **108**, 3123.

[36] F. J. Schmitz, S. P. Gunasekera, G. Yalamanchili, M. B. Hossain and D. van der Helm, Tedanolide: a potent cytotoxic macrolide from the Caribbean sponge *Tedania ignis*. *J. Am. Chem. Soc.*, 1984, **106**, 7251.

[37] D. O'Hagan, Evolution of the polyketide metabolites. *Chem. Br.*, 1990, **26**, 246.

[38] The stereochemistry of tedanolide was incorrectly drawn in reference [36]. See N. Fusetani, T. Sugawara, S. Matsunga and H. Hirota, *J. Org. Chem.*, 1991, **56**, 4971.

[39] Y. Kashman, A. Groweiss and U. Shmueli, Latrunculin, a new 2-thiazolidinone macrolide from the marine sponge *Latrunculia magnifica*. *Tetrahedron Lett.*, 1980, **21**, 3629.

[40] P. Crews, Y. Kakou and E. Quinoa, Mycothiazole, a polyketide heterocycle from a marine sponge. *J. Am. Chem. Soc.*, 1988, **110**, 4365.

Index